Advance Praise for
Lyme Disease, second edition

"An outstanding book on the history, diagnosis, treatment, preventive measures, and natural course of Lyme disease. Alan Barbour's *Lyme Disease* is comprehensive and understandable—an informative and much-needed book both for medical and scientific professionals and for the thousands of people annually exposed to tick bites."

> JOHN F. ANDERSON, Distinguished Scientist, Emeritus,
> The Connecticut Agricultural Experiment Station

"If you are seeking reliable and accurate evidence-based information on Lyme disease, this is the first book to read. It is a welcome relief from the cacophony of false and misleading information being disseminated in the media, and thereby makes a significant contribution to public health. It will be a valuable reference resource on Lyme disease for many years to come."

> PHILLIP J. BAKER, Executive Director, American Lyme
> Disease Foundation

"Science is complex—infections and the immune response to them particularly so—and complexity sows confusion. Misinterpretations of the biology of Lyme disease have created an air of mystery, making it possible for fanciful notions to grow up around this 'great imitator.' In this charming volume, Dr. Barbour, a pioneer of this field, provides a remarkably clear yet subtle description of these complexities, readable by any nonscientist who truly wants to understand this fascinating infection. A must read!"

> JOHN J. HALPERIN, Atlantic Neuroscience Institute,
> Overlook Medical Center and Icahn School of Medicine
> at Mount Sinai

"Want to really understand Lyme disease? Here is an engaging, conversational review that is presented with empathy. It is clear, thorough, nuanced, and evidence based. Grounded in the principles of good medicine, this book will be a valuable resource for health care providers and patients alike."

> BARBARA J. B. JOHNSON, retired from the Centers for
> Disease Control and Prevention

"Alan Barbour has been at the forefront of Lyme disease research for over 40 years. His book is an invaluable resource for everyone concerned about the disease."

LYME DISEASE

A Johns Hopkins Press Health Book

LYME DISEASE

Why It's Spreading,
How It Makes You Sick,
and What to Do about It

ALAN G. BARBOUR, MD

JOHNS HOPKINS UNIVERSITY PRESS
Baltimore

616.9246 BAR

Note to the reader. This book is not meant to substitute for medical care of people with Lyme disease or another infection transmitted by deer ticks, and treatment should not be based solely on its contents. Instead, treatment must be developed in a dialogue between the individual and his or her physician. Our book has been written to help with that dialogue.

Drug dosage: The author and publisher have made reasonable efforts to determine that the selection of drugs discussed in this text conform to the practices of the general medical community. The medications described do not necessarily have specific approval by the U.S. Food and Drug Administration for use in the diseases for which they are recommended. In view of ongoing research, changes in governmental regulation, and the constant flow of information relating to drug therapy and drug reactions, the reader is urged to check the package insert of each drug for any change in indications and dosage and for warnings and precautions. This is particularly important when the recommended agent is a new and/or infrequently used drug.

© 1996, 2015 Johns Hopkins University Press
All rights reserved. Published 2015. First edition published as *Lyme Disease: The Cause, the Cure, the Controversy*, 1996.
Printed in the United States of America on acid-free paper
9 8 7 6 5 4 3 2 1

Johns Hopkins University Press
2715 North Charles Street
Baltimore, Maryland 21218-4363
www.press.jhu.edu

Library of Congress Cataloging-in-Publication Data

Barbour, Alan G., 1946–
 Lyme disease : why it's spreading, how it makes you sick, and what to do about it /
Alan G. Barbour, MD.
     pages cm. — (A Johns Hopkins press health book)
 Includes bibliographical references and index.
 ISBN 978-1-4214-1720-2 (hardcover : alk. paper)—ISBN 978-1-4214-1721-9 (pbk. : alk. paper)—ISBN
978-1-4214-1722-6 (electronic)—ISBN 1-4214-1720-0 (hardcover : alk. paper)—ISBN 1-4214-1721-9
(pbk. : alk. paper)—ISBN 1-4214-1722-7 (electronic)   1. Lyme disease—Popular works.   I. Title.
 RC155.5.B38 2015
 616.9'246—dc23 2014036888

A catalog record for this book is available from the British Library.

Special discounts are available for bulk purchases of this book. For more information, please contact Special Sales at 410-516-6936 or specialsales@press.jhu.edu.

Johns Hopkins University Press uses environmentally friendly book materials, including recycled text paper that is composed of at least 30 percent post-consumer waste, whenever possible.

Contents

Preface vii

Introduction. Three Case Histories and a Parable 1

1 Early Infection and the Immune Response 12

2 Late Infection and Its Complications 27

3 The Pathogen, Its Vector, and Its Reservoirs 42

4 The Ecology of Lyme Disease 60

5 Approach to Diagnosis 82

6 Laboratory Tests: The Basics 101

7 Putting Laboratory Testing in Its Place 124

8 Antibiotics and Lyme Disease: The Basics 140

9 Putting Antibiotics to Use 159

10 After Antibiotic Therapy Ends 182

11 Deer Ticks Transmit Other Diseases 198

12 Preventing Lyme Disease: Community-Wide Measures 219

13 Preventing Lyme Disease: Personal and Household Protection 248

Trusted Internet Sites 271

Notes 275

Glossary 301

Index 307

Preface

Some readers may be wondering whether this is a new edition of my 1996 book or a new book. A more pressing question might be why a new book or a new edition now? One reason is that large portions of the original book are admittedly out of date (only about 10 percent of the first book's text was retained in this edition). In the interval since 1996, Lyme disease has continued to rise in frequency and to spread its reach in North America, Europe, and Asia. A human vaccine has come and gone from the market. There are new diagnostic tests, greater appreciation of the long-term consequences of the infection for some patients, and better understanding of what could be effective measures to control Lyme disease. A second reason is the increasing public health importance of other infections transmitted by the same ticks that carry the Lyme disease microbe. These include a protozoan related to the cause of malaria, a deadly virus of the brain, and other types of pathogenic bacteria, one of which was only recognized as a cause of human disease in the last year. These other diseases merit increased attention.

The title—*Lyme Disease*—is the same, but the new subtitle diverges from *The Cause, the Cure, the Controversy*. While acknowledging the continuing divide on what the definition of Lyme disease should be and how the disease should be diagnosed and treated, I concluded that to do justice to each point of controversy, with presentations of all sides, would consume most of the book's allotted pages. I touch upon the controversy as needed and provide readers with sources that seem to be fairly representative of, if not the latest word on, the different positions.

But the viewpoint of this book is essentially as I described for the first book: in cases where there are controversies, I give the opinion or rec-ommendation that is best supported by the available scientific (i.e., evidence-based) data. As the 1996 book preface stated: "Some unproven theories and ideas are very appealing, but they are difficult to recom-mend if the only basis for their acceptance at this point is anecdotal. Compelling stories, many of them from patients, should be taken seri-ously: they may point the direction to new concepts of disease and treatment. But ultimately those holding views that are contrary to what has been shown scientifically must put their own theories to the test."

In the interval since the first book, my studies of Lyme disease, the bacteria that causes it and other diseases, the wildlife that carry them, and the ticks that transmit them have been supported by grants from the National Institute of Allergy and Infectious Diseases of the Na-tional Institutes of Health, the Centers for Disease Control and Pre-vention, the National Research Fund for Tick-Borne Diseases, and the Ellison Medical Foundation's Senior Scholar Award in Global Infec-tious Diseases. The research has also been partially supported by royal-ties from licensing of patents for a vaccine against Lyme disease for hu-mans and dogs and for a component of a diagnostic assay for Lyme disease in humans and dogs. At pertinent places in the book, I disclose when consideration of a vaccine or a diagnostic assay might be a con-flict of interest on those grounds.

Since publication of the first book, I have not received compensation or research funding from consulting for pharmaceutical, vaccine, diag-nostics, or insurance companies, for serving as an expert witness, or for performing diagnostic assays. While I have practiced medicine in the past and have cared for patients with a variety of infectious and other disorders, I do not do so now. I am a full-time university faculty mem-ber who continues to carry out research projects on Lyme disease and relapsing fever that date to 1981, to teach microbiology and clinical in-fectious diseases to medical students and residents, to correspond with colleagues in these fields, to attend scientific and medical meetings, to be a peer reviewer on grant proposals and submitted manuscripts, and to write professional journal papers, review articles, and medical text-book chapters on these subjects.

The list of all the people I could acknowledge for their contributions to the field and to the successes of my laboratory's endeavors, as well as for their guidance and advice, would be at least twice the length of the first book's already extensive list. But I cannot neglect to recognize again a handful of them. These are the late Wilhelm "Willy" Burgdorfer, for his mastery of medical entomology and his acceptance of a junior scientist as an equal in a discovery of a lifetime, and Durland Fish, Allen Steere, and Gary Wormser, not only for their past and ongoing contributions but also for their courage to call it as they see it. I also thank Joe Piesman of Colorado State University and Marc Dolan of the Centers for Disease Control and Prevention for providing comments on chapters 12 and 13 as well as Melanie Mallon for her sympathetic and expert copy-editing of the manuscript into final shape.

This book is dedicated to the memories of George Barbour, whose aspiration to be a scientist was thwarted by providence, but who passed on his open-minded curiosity to his son; of John Swanson, who encouraged my pursuit of a long-neglected disease, relapsing fever, the upshot of which was codiscovery of Lyme disease's cause; and of Paul Lavoie and Ed Masters, two private practitioners whom I did not always agree with but who were inspirational in their passion to learn from their patients and thereby help others. Special thanks goes to Jacqueline Wehmueller, my long-patient editor at Johns Hopkins University Press, for keeping the flame for a revision going even when faint and for then fostering the growth of what in the end is a new book. Finally, I could not have done any of this over the past four decades without the support in so many ways of my wife, Ann.

LYME DISEASE

Introduction
Three Case Histories and a Parable

■ **Case A.** In Stockholm of 1909, the October monthly meeting of the Dermatological Society was held on the 28th. As was custom, brief summaries of the
presentations were subsequently published in German in a medical journal.[1] The
last entry for that day's meeting recorded that "Afzelius erwähnt ein von *Ixodes
reduvius* wahrscheinlich hervorgerufenes Erythema migrans bein einer älteren
Frau." The Afzelius who "mentioned" (*erwähnt*) was Dr. Arvid Afzelius, a
52-year-old Swedish dermatologist and author of a textbook on skin diseases.[2]
Dr. Afzelius's patient was "an older woman" (*einer älteren Frau*), and the "erythema
migrans" condition presumably was uncommon enough to merit mention at a
medical meeting. According to the article, Afzelius thought it was "probably
caused" (*wahrscheinlich hervorgerufenes*) by a tick, "*Ixodes reduvius.*" This tick is
now named *Ixodes ricinus*, commonly known as the sheep tick and found
throughout much of Europe, including Sweden.

What was this erythema migrans that was of interest to the roomful of
dermatologists? As was common for a new skin disorder, Afzelius assigned
a Latin name, the literal translation of which is "migrating redness." But
it was not migrating in the sense of decamping from one place and moving to another, as in migrations of birds or refugees. The redness did not
move from one part of the body to another part. A closer rendering is
"advancing redness": a ringlike rash whose border of redness expands as
its center clears. We learn this from a more definitive account of erythema
migrans by Afzelius in 1921, as well as by others who reported on the condition from different parts of Europe over the succeeding years. As time

went on, erythema migrans in Europe was linked to another tick-associated disorder, one of the nervous system, and to a longer-lasting skin disease. In the 1940s and 1950s, several papers proposed that erythema migrans and its associated conditions were an infection.[3] By the 1960s, treatment of erythema migrans with penicillin was a common practice in Europe. But this therapy was not based on knowledge of what exactly caused the infection. It was empirical—in other words, based on an informed hunch (and as it turned out, a good one). The actual cause remained unknown for several more years, and it was discovered first on another continent.

■ **Case B.** On Monday, July 1, 1974, a 40-year-old woman living on Shelter Island, New York, visited the island's family medicine doctor for fever and severe headache. These symptoms were followed by paralysis of a nerve for the muscles on one side of her face, causing a droop on that side. After two weeks, this illness resolved without treatment. In June 1975 she had severe pain and swelling of the right knee, which also improved after a few weeks without treatment. But there were recurrences of swelling and pain of the right knee, with fluid buildup in the joint in August 1976 and then again in 1977 and 1978. In September 1978, her physician prescribed tetracycline by mouth for two weeks, and she had no further episodes of knee arthritis over the four years of followup. In 1982 the doctor sent the patient's serum specimen, which he had kept in storage in his freezer for those years, to the Rocky Mountain Laboratories in Montana.

At that outpost of the National Institutes of Health in fall 1981, a hitherto unknown bacterium had been discovered in deer ticks that had been collected on Shelter Island earlier in the year. The Montana scientists found that the patient's serum had high levels of antibodies to the new microorganism, as did the sera of several other of the doctor's Shelter Island patients with similar illnesses.[4] This new bacterium was initially called the *Ixodes dammini* spirochete, after the deer tick's formal name at the time and the class of bacteria to which it belonged.

This woman's illness began two years before a July 18, 1976, *New York Times* article on a new form of arthritis in and around Lyme, Connecticut.[5]

Lyme is directly across the Long Island Sound from Shelter Island. The newspaper article recounted the investigations of Drs. Allen Steere and Stephen Malawista at Yale University, Dr. David Syndman at Connecticut's health department, and other colleagues. These investigators' 1977 account of their findings in a medical journal began, "In November 1975 a mother from Old Lyme, Connecticut, informed the State Health Department that 12 children from that small community of 5,000, 4 of whom lived close together on the same road, had a disease diagnosed as juvenile rheumatoid arthritis."[6] Observed features of the new disease included onsets of the arthritis in late summer or early fall, peak age incidences in late childhood and early middle age, equally affecting males and females, lack of evidence of person-to-person transmission, a tendency for the arthritis to come and go, rare involvement of the hands and feet, and the frequent report of a peculiar ringlike skin rash that preceded the onset of arthritis. All the cases in the original investigation lived in heavily wooded areas and not the more densely populated town centers.

The 1976 newspaper article reported that the investigating team "found a common pattern among the cases that suggests the disease is a form of arthritis caused by a virus carried by an insect or other biting arthropod such as a tick." An arthropod-transmitted virus was considered the most likely cause, and major efforts were directed at discovering what the virus was. These proved to be unfruitful, though, and the major advances of the next few years were fuller characterization of the skin rash in its typical and atypical forms and establishment of the strong association between what was now called Lyme disease and exposure to ticks, specifically the deer tick of the area, called *Ixodes dammini* then and later named *Ixodes scapularis*. Besides the skin rash and the arthritis, the expanded definition of Lyme disease also included some neurologic disorders, such as facial nerve paralysis; some abnormalities in the conduction of nerve impulses in the heart; and inflammation of the brain. Most patients were not treated with antibiotics because Lyme disease was still thought to be caused by a virus, against which antibiotics would be ineffective. As a consequence, several of the individuals in and around Lyme progressed through different stages of the disease, from an early stage in which there was a skin rash and symptoms like fever and fatigue and then to nerve disorders and pain in many joints and finally to joint pain and swelling in one or two large joints, like the knee or hip.

The first of two turning points was the recognition that the skin rash occurring around Lyme and other places in the Northeast was similar to the erythema migrans that had been reported in Europe since 1909 (Case A). By the 1960s, the standard of care in Europe for patients with the skin rash or the later neurologic disorder was administration of an antibiotic, usually penicillin. This connection between the two diseases on each side of the Atlantic soon led to controlled clinical trials of antibiotics in the United States. These showed a clear benefit of antibiotic treatment over no treatment for both the early form of Lyme disease, such as when there is a rash and fever, and the later forms, when there is neurologic, heart, or joint involvement.[7] Although there remains controversy about who should get antibiotics, in what form, and for how long they should be given, there is no doubt that the first line of treatment of Lyme disease is an antibiotic. In addition to the clinical trial results, strong evidence of the value of antibiotics has been the large decline in the occurrences of patients like Case B or the children in Lyme with swollen, painful joints.

The second turning point was the aforementioned discovery of a spirochete type of bacteria in the ticks of Shelter Island, the isolation of the bacteria in the laboratory, and the demonstration that the Shelter Island patients who had Lyme disease had antibodies to it. This discovery was made about the same time that the effectiveness of antibiotics was reported, so it was easier for prevailing medical opinion to accept that a bacterium was the cause and not a virus. Soon followed the recovery of the bacterium, henceforth named *Borrelia burgdorferi*, from patients with different forms of Lyme disease, and wide application of diagnostic tests for antibodies to *B. burgdorferi*.[8] These showed that patients with similar diseases in north-central states and in Northern California, as well as in the northeastern United States, also had *B. burgdorferi* infection.

Once *B. burgdorferi* was discovered in North America, medical researchers in Europe applied the method and isolated it and related bacteria from patients with erythema migrans and the associated neurologic disorder, as well as from the *Ixodes* species of ticks that were associated with the infection on that continent. One of the new related species of *Borrelia* that were discovered in Europe was given the species name *afzelii* in honor of Arvid Afzelius.

■ **Case C.** An otherwise healthy 38-year-old man living in north Texas began to have severe pain in his leg in late November. After initial diagnoses of an inflamed vein or muscle strain turned out to be inaccurate, he sought an answer at a large referral medical group in Minnesota. In the course of the medical workup, the physicians there performed a blood test for Lyme disease. The result came back negative, and the attending physicians could not offer a definitive diagnosis. Still seeking an answer for the pain and a feeling of exhaustion, he saw a Dallas-area neurologist who specialized in Lyme disease. This physician obtained another blood sample and sent one portion to a national reference laboratory and another portion to a Northern California private laboratory, which offered a more limited set of diagnostic tests but which was a favorite of some doctors. In both laboratories, the first test, known as an ELISA, for enzyme-linked immunosorbent assay, was interpreted as positive. The Western blot test, the follow-up procedure to a positive ELISA result, was then run at both laboratories. The result was negative at the national laboratory but positive at the special laboratory in California. His doctor in Texas diagnosed Lyme disease and started the patient on an oral antibiotic. But on the advice of a colleague, the patient sought a second opinion at a tick-borne diseases referral clinic in Westchester, New York. The Western blot test was repeated at the clinic's laboratory and was negative. The physician in New York told the patient that he did not have Lyme disease. The patient was unsure who to trust and continued to take the antibiotics prescribed in Texas. But by July of the next year, he concluded that he did not have Lyme disease and stopped the antibiotic for lack of benefit. The patient was quoted in a newspaper as saying, "It's been a hell of an emotional roller coaster." According to the article, he acknowledged that having the Lyme disease diagnosis for a while was a comfort.[9]

The third case history recalls a thought experiment, a working through in the mind of the consequences of a hypothesis. In this thought experiment, there is Earth as we know it and a so-called Twin Earth. Twin Earth is identical but for one difference. In the American philosopher Tyler

Burge's example, there is an individual on Earth, we will call Fred, who has had arthritis of the knee joint and some other joints.[10] Up to this point, Fred's disorder corresponds to the accounts of "arthritis" in standard medical textbooks on Earth. But one day, Fred feels pain in his thigh muscle and thinks, "I have arthritis of my thigh." He sees his physician with this complaint. The doctor does not deny Fred's perception of pain in the thigh muscle but explains that it cannot be arthritis per se, because that term only applies to a disorder of the joints. Fred defers to the physician's expertise and accepts that his belief about arthritis of the thigh had been false, but for him there is still something wrong with his thigh.

On Twin Earth, in this telling, both the medical and lay meaning of "arthritis" encompasses not just disease of the joints but disease of muscles and tendons as well. Our Twin Fred, with exactly the same experience of pain in his thigh and the same history of painful joints—that is, the intrinsic facts are the same—receives a different reception when he says, "I have arthritis of my thigh," to the Twin Earth doctor. On this other world, it was not a false belief. The Twin Earth doctor and Twin Fred are in agreement. Whether Twin Fred can obtain any relief for his painful thigh remains to be seen, but he does have a diagnosis for the ailment: arthritis.

With some latitude this twin world analogy applies here, with Lyme disease taking the place of arthritis. Here is my Earth version of Lyme disease. It is a bacterial infection, as strep throat and gonorrhea are bacterial infections. Strep throat is caused only by the bacterial species *Streptococcus pyogenes*; gonorrhea is caused only by *Neisseria gonorrhoeae*; and Lyme disease in North America is caused by *Borrelia burgdorferi*. Strep throat bacteria and gonorrhea bacteria are spread from one person to another directly. But almost the only way to become infected with *B. burgdorferi* is through the bite of a tick. And not just by any type of tick. Many types of ticks commonly bite people yet are incapable of transmitting Lyme disease. Or the tick is both capable and infected but is not one that people would encounter. Or the right sort of tick is there, but *B. burgdorferi* is absent from the wild mammals and birds in the area, so the ticks have nothing to pass on. The conditions have to be "just so" for there to be enough infected ticks of the human-biting variety to be a significant threat for people. This means that the risk of getting infected varies greatly across North America and according to the time of year. In some areas, the risk is nonexistent at any time of year. In others, it may be high, but only during certain months.

Let's assume that the bacteria are present in human-biting ticks and that certain activities, like gardening, playing in the nearby woods, or hiking, have put a resident of the area at higher risk. That meets an indispensable criterion of a diagnosis: acquiring the infection is possible, maybe even probable. Let's go on to imagine that this person presents to the doctor with an illness that in its cumulative features could be Lyme disease, which is perhaps not at the top of the list to start but is a serious contender. And, further, the interval between the tick exposure and the illness is measured in days or weeks, months at the outside, but not in years. (Details of *B. burgdorferi* infection in its different forms and stages are considered later.) Commonly, there is a combination of symptoms that are nonspecific, like fever, muscle aches, and fatigue, and those that more narrowly point to Lyme disease, for instance, a rash such as erythema migrans or a certain disorder of the nerves.

If the clinical findings together with an actual exposure risk incriminate Lyme disease, and it is still in its early phase, the physician may prescribe antibiotics by mouth. The prescription might be accompanied by an order for a blood test for the presence of antibodies to *B. burgdorferi*. The physician knows that the result may be negative if antibodies in response to the infection have not yet appeared. On the other hand, they might be present at an early stage, and in any case, the result provides a baseline for a follow-up test. If the infection is further advanced in its course or not so typical in its features, and there is not a reason, such as a heart or brain disorder, for immediate treatment, the physician orders the same blood test and waits for a day or two for the result. Less commonly there is an attempt to recover *B. burgdorferi* from the blood or a skin biopsy, either by culture in the laboratory or by the PCR assay, which detects the DNA of a pathogen (chapter 6). Isolating the organism or its DNA from the patient would be the strongest evidence that it was a cause of the current illness, as is the case for a diagnosis of strep throat or gonorrhea. But timing is everything here; once the infection is in its later phase, recovery of the microbe by these means is uncommon.

The ticks that transmit *B. burgdorferi* may also transmit other pathogens, so these infections alone or in combination with Lyme disease are considered. For the question of Lyme disease itself, however, if the physician concludes that an infection with *B. burgdorferi* is ongoing, an antibiotic is the treatment. It is possible that the infection would resolve on its own or be of no consequence to health, but the standard of care correctly calls for antibiotic therapy. If the diagnosis of active *B. burgdorferi* infection is accurate and the appropriate antibiotic prescribed and taken as directed, the great majority of patients benefit. Most will be cured. As with a strep throat or gonorrhea, some patients may not be cured, and this prompts another round of an antibiotic, perhaps a different one. A minority of patients may be technically cured of an active infection, as we usually define it, but continue to be ill. This may not be in the way they originally presented, but they still do not feel how they felt before getting sick. I consider this phenomenon in more depth later (chapter 10). Suffice it to say here that what has been called a "post–Lyme disease syndrome" is a sequela of *B. burgdorferi* infection by definition.

A Twin Earth book on Lyme disease could have several points in common with the Earth account of Lyme disease. If there is consensus that the disease is caused by a bacterium—as opposed to a virus, unspecified toxin, or malevolent force—and is almost exclusively acquired from a tick bite, there can be agreement on ways to reduce the risk of infection to begin with and perhaps on the efficacy of short courses of antibiotic treatments for cases that all parties agree is early Lyme disease. But from there, the definitions and views diverge. What I call the main alternative view of Lyme disease holds the following tenets in opposition to the positions of most medical and public health experts: (1) the incidence and geographic distribution of Lyme disease in North America and Europe is greater than what governments estimate, even after official adjustments for nonreporting; (2) the list of possible symptoms or maladies qualifying as Lyme disease is longer than what traditional medical textbooks say; (3) coinfections of *B. burgdorferi* with other microbes are more common than is generally appreciated; (4) the standard laboratory tests are either inadequate to begin with or their results are interpreted too strictly; and (5) persistence or recurrence of symptoms, with perhaps the addition of new symptoms, after standard antibiotic treatment—that which most insurance companies would allow reimbursement for—is due to presence of live surviving bacteria, thus justifying longer antibiotic treatments, sometimes intravenously, sometimes in combinations, that are measured in months.

Many books have been written from alternative perspectives and are widely available for those seeking other opinions and options. At least two documentary movies with this point of view have been released.* In both North America and Europe, websites and social media outlets of several patient advocacy groups, some with national scope and some more regionally oriented, unambiguously represent alternative views. For the curious or unsatisfied, these groups provide much information on their positions and recommendations. There is even a professional medical organization, the International Lyme and Associated Diseases Society (ILADS), that to my view largely espouses the alternative conception of Lyme disease. The society's website would be a good jumping-off point for those interested. The rapid multiplication of so-called open access scholarly journals, which are often published by for-profit companies and in some cases effectively provide self-publishing with little or no rigorous peer review,[11] has meant that journal articles representing the alternative view (and for better or worse, the traditional view as well) are more numerous in freely accessible medical literature than they used to be. A patient or concerned family member sifting through all the sources and varieties of information, some diametrically opposed, can understandably feel unsure what to believe or trust.

I understand the frustrations often felt by people with unexplained disorders and disabilities. Over the years, I have listened to many of their stories. I also understand the draw of the alternative view and the broadened definition of Lyme disease. If the choice is between "We're not sure

---

* *Under Our Skin* (Open Eye Pictures) in 2008 and *Under the Eightball* (Andalucian Dogs) in 2009.

what you have" and "You've got Lyme disease," who would not find some relief in hearing the doctor give choice number two, even if the relief is short-lived? Fred was still left with an unexplained thigh pain on "Earth." In my opinion, there has not been enough attention paid by government agencies and academic institutions to the chronic Lyme disease phenomenon, which is as much social as biological.[12] But this book does not provide either a roadmap to or an interpretation of the alternative views, except in some places in the book where some comment is called for. This is a book mainly from the Earth perspective, with acknowledgements of the issues the Twin Earth view has productively raised.

What follows are chapters on Lyme disease in its early stage (chapter 1) and in its later stage (chapter 2); the biology of the microbe, the tick that transmits it, and the animals that carry it (chapter 3); the ecology of Lyme disease (chapter 4); how diagnosis is made (chapter 5); the basics of laboratory tests (chapter 6); how test results are interpreted (chapter 7); antibiotic fundamentals (chapter 8); Lyme disease treatments (chapter 9); what happens when the patient still feels ill after treatment (chapter 10); other infections transmitted by the same type of tick (chapter 11); disease prevention at the community-wide level (chapter 12); and personal protection for individuals and households (chapter 13). Rather than give precise recommendations on diagnostics or regimens, which may be superseded in the next months or years, I emphasize general principles that would likely govern new procedures, therapies, and preventive maneuvers as well as current ones. The patient histories are abstracted from real cases, not fictionalized ones. Although I cite selected literature sources, such as for the case histories and particular studies, I have not used exhaustive citations. A list of websites in the "Trusted Internet Sites" section near the end of this book provides entrée for those seeking additional information. The back pages also contain a glossary of selected terms as they are applicable in the context of Lyme disease and this book.

I use the term "Lyme disease" instead of the alternative name "Lyme borreliosis" throughout. In my opinion, "Lyme borreliosis" is preferable and may one day supplant "Lyme disease" as the official name. "Borreliosis" more clearly denotes the infectious basis of the disorder, but frankly, it is not nearly as well known among the public and the media. Most state governments and the federal government in the United States still record cases of "Lyme disease." The relevant diagnosis code under the widely adopted international code of diseases and procedures, the ICD-10, of the World Health Organization is for "Lyme disease."

The "Lyme" part of the disease's name will probably stick, even though it is an accident of history rather than the consensus of a panel of experts. Unfortunately, many people think it was named after a person, maybe a "Dr. Lyme" somewhere, so it ends up on people's tongues as "Lyme's disease." Recognition of Lyme as a place name, at least, can lead one to the medical detective story in the 1970s or to a place on the map that is ground zero for the high-risk area for infection in the northeastern United States. A search for a "Dr. Lyme" in this context will be fruitless.

The name is ironic in that what we refer to as Lyme disease was described in many of its aspects by European physicians, like Arvid Afzelius of Case A, several decades before the U.S. outbreak of arthritis in children was reported. Although this tick-borne infectious disease is as common and long established in Europe as it is in North America, I mainly had North American readers in mind while writing. The book points out some of the distinctions between the infections on each side of the Atlantic. One of those differences is the much lower frequency of arthritis as a complication of the infection in Europe than in North America. This explains in part why "Lyme arthritis" took its name from a Connecticut shore town founded in the seventeenth century rather than the town's namesake in England: Lyme Regis on the Dorset coast, mentioned in the Domesday Book of 1086.

Chapter 1

Early Infection and the Immune Response

■ **Case D.** A 73-year-old man living in Connecticut noted a small red patch on his left thigh. He thought this was a reaction to an insect's bite, but the red patch enlarged outwardly, with clearing of the redness in the center, over the next month. When he was seen by his physician, the rash measured 3 by 6 inches (8 by 15 centimeters) and was accompanied by discomfort and weakness in his left leg. He was treated with the antibiotic erythromycin by mouth (oral) and was asymptomatic two days later. At a three-month follow-up appointment, he remained symptom-free.[1]

■ **Case E.** A 32-year-old man hiked in a rural area north of San Francisco. The next day, he discovered on his thigh an embedded tick, which he removed and saved. One week later, he noticed a red, slightly itchy ring on the skin that surrounded the site where the tick had attached. Four weeks after he removed the tick, he was seen by his physician, who recorded a red ring, 3 by 5 inches (8 by 12 centimeters). By eight weeks, the ring had become fainter, but it had expanded over the leg to 14 by 14 inches (35 by 35 centimeters). He was treated with oral tetracycline and observed that the ring had disappeared by three days later. The tick was identified as the western black-legged tick, whose formal name is *Ixodes pacificus*.[2]

■ **Case F.** On August 3, 1982, a 70-year-old woman who was a summer resident on an island off the New York mainland noted a red flat patch of skin measuring 1 by 3 inches (2 by 7 centimeters) on her inner thigh and sought medical attention. She also had a body temperature of 101.7°F (38.7°C), muscle aches of the back and neck, and fatigue. She did not conclusively recall a tick bite. A blood sample was obtained on August 10. The patient's physician prescribed oral penicillin. Within 10 days of starting the antibiotic, the patient was asymptomatic, and the rash had cleared. A second blood sample was obtained on September 22. A portion of the blood specimen from August 10, one week into the illness, was added to a tube of culture medium. The bacterium *Borrelia burgdorferi* grew out of that culture in the laboratory. Although the August 10 blood sample did not have antibodies to the organism, antibodies to *B. burgdorferi* were detected in the September 22 specimen.[3]

How Does Lyme Disease Start?

The people in each of these case histories had encountered a tick, but only Case E was definitely aware of it. Case D thought he had an insect bite, and Case F wasn't sure. But both D and F lived in areas where deer ticks are commonly encountered and where Lyme disease is a known risk for residents and visitors. So, there was the opportunity for a tick bite, if not recollection of one. Up to half the ticks in some areas, for example, eastern Connecticut and Long Island, New York, are infected. In other areas where Lyme disease occurs, such as Northern California, the percentage of ticks with the spirochete may be as low as 2 percent. (Chapter 4 considers why the chance of infection varies from one place to another and what puts an individual at greater risk.)

For each of the three people, the events of the illness started when a tick latched on to clothing or exposed skin from its perched position on some high grass or bushes. The type of tick that commonly transmits Lyme disease to humans is about the size of a sesame seed, so it can understandably be overlooked. (Differences among ticks are considered in chapter 3.) If the tick is on the clothing, it crawls around until it finds skin. Then the tick attaches and buries its rasplike mouth into the upper layer of skin. It fastens itself tight in place with a cement-like material it secretes. The tick then excavates a small cavity in the skin and begins feeding on a mixture of blood and tissue fluid that accumulates there, injecting saliva through its mouth as it feeds. Potent substances in the saliva inhibit the blood from clotting and thwart some of the human body's first-line defenses against the tick. These substances diminish the reaction against the tick's presence in the skin, so it is less likely that there will be pain or itchiness, which would draw our attention.

With the tick firmly embedded and blood entering its intestine, the spirochetes are stimulated to leave their residence in the tick's gut. They move into its fluid-filled body cavity, and then swim to the saliva glands near the mouth. From there they leave with the stream of injected saliva fluid, thereby entering the human skin. This migration through the tick begins about one or two days after the tick first attaches itself.

Often people will see the tick while it is feeding and remove it. If they do this during the first day or two after the tick embeds, the chances of getting Lyme disease are lower than if they leave it undisturbed longer. Case D noted and extracted the tick the day after exposure, but it was too late in his case. The proverbial horse had left the barn. In other instances, the tick falls off on its own after its blood meal ends, in three to seven days. After completion of feeding, the tick is distended, full of blood, a state called "engorged." The period for complete feeding is time enough for transmission of spirochetes to occur. But even under optimum circumstances for transmission, infection is not inevitable. After the tick takes in blood, the spirochetes begin to multiply to large numbers— ten thousand or so, but only a small fraction of these move from the intestine to the salivary glands. And by the time some make it that far, there may be fewer than one hundred that are in place for the jump to the human. In fact, the odds of a successful transmission from an infected tick to a human host are less than 5 percent.[4]

A residual indication of a tick bite may be a small red bump that lasts for a few days. Some people have allergic reactions to the ticks themselves, in which case the area of redness around the bite may be larger, and it may itch. This is not an infection per se, even though there is redness and swelling—hallmarks of inflammation, a word derived from the Latin word *inflammare*, meaning "to set on fire." Inflammation occurs with infection, but the type of inflammation that occurs in this instance is similar to a skin hypersensitivity to a cosmetic or to certain metals in jewelry. This type of allergic reaction to ticks can develop after an earlier bite by the same type of tick.

Sometimes people develop a bacterial infection at the tick bite, but this infection is not Lyme disease. Such infections can occur after any break or cut in the skin. They are caused not by spirochetes but by other bacteria, such as those that cause strep throat or the staph bacteria of boils and abscesses. In these cases, the tick cannot be blamed for carrying the disease, because the bacteria are usually on the person's skin when the tick arrives. But by breaking the protective surface of the skin, the tick bite allows the bacteria to invade the person's deeper tissue. The infection causes the skin around the tick bite to become inflamed and often very tender. The size of the inflamed area may increase substantially within a matter of hours. The medical term for this type of infection is cellulitis. Sometimes cellulitis is confused with early Lyme disease. It is important to distinguish the two, because the antibiotics used for treatment are generally different (chapter 9).

Let's assume there was neither an allergic reaction to the tick itself nor cellulitis. If we could look into the skin around where the tick had fed, we would see a few spirochetes in the spaces between the skin cells. Spirochetes are long, wavy bacteria that have the ability to move and to change direction. They travel as a snake does when it moves over the ground. And like a snake, they can fit into tight spaces, such as those between human skin cells. Spirochetes require simple sugars, fats, and other substances for energy and growth. Because they cannot make these on their own, they have to get them from the host's surrounding tissues or blood. These spirochetes are parasites; they take from their host but do not give anything in return. As they divide and multiply within one area of the host's body, they begin to compete among themselves for food resources. Then, in order to survive, they move out from their starting place to search for new sources of food.

The First Response to Infection

While the spirochetes are multiplying and feeding, the human who was infected by the tick begins to respond to their presence. The response occurs automatically, with neither the need for nor the possibility of conscious direction. This defense is the immune response that occurs when human bodies encounter viruses, bacteria, and other microorganisms. An immune response also occurs when a person gets a vaccination, receives an organ transplant from someone who is not closely related, or develops an allergy.

The earliest immune responses to a microorganism's presence tend to be general—that is, their mechanisms protect people against a variety of microscopic threats. Humans and other vertebrates have this first line of defense—called innate immunity—in common with simpler organisms, like flies and earthworms. One example, lysozyme, an enzyme in the tear fluid, kills—"lyses"—certain types of bacteria before they begin to multiply on the surface of the eye. Certain types of white cells in the blood and in the tissues can also respond quickly to a microbe's presence. These white cells are called "phagocytes," a name derived from the Greek word *phagein*, for "eat." The phagocytic type of white cells literally eat the bacteria they encounter or seek out. They envelop the bacterial cell and draw it into the white cell's interior, where it can be destroyed by enzymes and other substances the white cell makes. Some other proteins, which are found in the blood, bind to bacteria and either kill them directly or make them better targets for phagocytic cells. Although most bacteria we encounter in our daily living are destroyed by these mechanisms if they get in the blood—and this happens not infrequently when we brush our teeth or have a bowel movement—the Lyme disease spirochetes can avoid both recognition by white cells and binding by on-guard blood proteins.

As the infection progresses, humans and other animals respond in another nonspecific way to the invasion. The presence of the bacteria or viruses directly or indirectly stimulates the production of hormone-like substances called cytokines, which act as chemical messengers between cells. By releasing a cytokine, a cell can influence the condition or behavior of another cell. The cytokine's effect can be local—for example, it can take place between adjacent cells—or it can be long distance, for instance, when cytokines enter the blood and are distributed throughout the body. One cytokine may turn on selected activities in the target cell, while another turns off a different set of activities. Most of the effects of cytokines are on white blood cells and other parts of the immune system, but they can also affect cells of the nervous system. One cytokine, called interleukin-1, causes blood vessels to dilate and the body's temperature to rise. Another cytokine, interferon-gamma, when administered as a medication has side effects of fever, headache, muscle aches, and fatigue. Cytokines and similar types of substances account for many of the general symptoms of early Lyme disease, like fever and muscle pains.

During the Lyme disease infection, the spirochetes seldom if ever become very abundant in the skin or elsewhere in the body (as we will see, this is one reason that Lyme disease is difficult to diagnose). They have an effect on the cells of humans and other mammals that is out of proportion to their numbers. Scientists have found that proteins on the surface of the spirochete cause human and animal cells to release cytokines in particularly large amounts. Whether spirochetes have anything to gain from stimulating cytokines is not clear. The local result, though, is a red rash like those of Cases D, E, and F. As occurs with cellulitis, the skin's redness is due to inflammation. The blood vessels dilate, and white blood cells move from the blood to the tissues in the area. The spirochetes, being mobile, may respond to all this activity by moving to tissue that is not yet inflamed. Thus, there are two good reasons for a spirochete's migration: one, to seek out food, and two, to escape the wave of inflammation that it has provoked.

The Skin Rash of Lyme Disease

Lyme disease spirochetes, unlike strep and staph bacteria, do not release toxins, that is, substances that can directly damage cells and tissues. The inflammatory response to the cells themselves produces the rash. As the spirochetes move away from the tick bite area, the inflammation response follows in their wake. The rash may be observed in as few as three days after exposure or, rarely, as long as a month after. In about half the patients, the rash appears within seven days. What starts as a small red spot grows, over the next several days to weeks, to become a large patch of red, which is usually circular or oval. The outer edge is distinct and well demarcated from the unaffected skin. Solitary rashes tend to be below the waist, on the trunk or legs, in adults. Particularly common locations for the rash are the groin, behind the knee, at the belt line, and in the armpit. In children the rashes are equally likely to be on the head and neck, the limbs, or the lower back. The abdomen is an uncommon place for a rash, probably because a tick there would have been more readily noted. In dark-skinned individuals, the rash may be more difficult to detect because there is less contrast between the rash and the surrounding skin. The usual criterion for a Lyme disease diagnosis is a rash of at least 2 inches (about 5 centimeters) at its widest, but a smaller rash does not rule out Lyme disease, especially if it is caught soon after the tick exposure.

Because there are fewer spirochetes left behind in the wake of the migration, the inner area of the rash is sometimes paler than the edge; the outer border may be bright red. The battle lines move to follow the advance of the spirochetes, leaving less inflammation in the center. The rash is called erythema migrans (EM). This is the name that Dr. Arvid Afzelius gave in 1909 (introduction). The ringlike rash is typical enough that physicians can often make the diagnosis of Lyme disease solely from the appearance of the rash, the patient's story, and a plausible risk of infection. Examples of EM are shown on the next page.

But this textbook ringlike rash is not invariably present. Less than half the patients with a rash of Lyme disease showed this pattern in one large study.[5] Besides the more solidly red patch, a variant on the expanding-ring pattern looks like a target, with two or three distinct rings surrounding either a pale or a red center. This bull's-eye pattern is also suggestive of Lyme disease, especially if, like the ring rash or large red patch, the target rash is either asymptomatic or only mildly itching or tingling. Occasionally, the center of the rash is the most intensely red portion and can be thickened or, rarely, covered with small blisters. On the other hand, 20 to 30 percent of patients with Lyme disease do not report having a rash at all. And, as we will see in chapter 11, an erythema migrans–like rash can also be a sign of a disease carried by another type of tick in the southern United States.

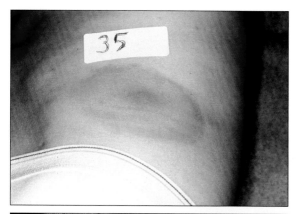

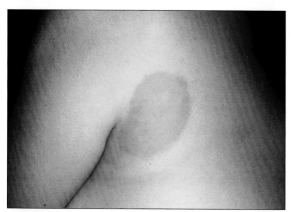

Solitary erythema migrans rashes of four patients with Lyme disease. The diameters of the rashes in the photographs are larger than most erythema migrans rashes at the time of examination, but they are representative of the variety of shapes and patterns of erythema migrans that may occur, even those of smaller size. Photographs by Gary Wormser, previously published in Alan G. Barbour and Durland Fish, "The Biological and Social Phenomenon of Lyme Disease," *Science* 260 (1993): 1610. Reprinted with permission from AAAS.

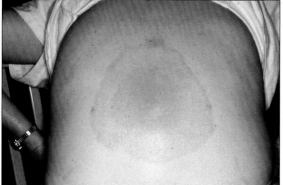

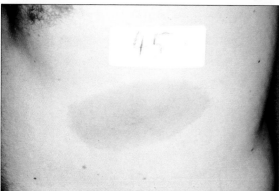

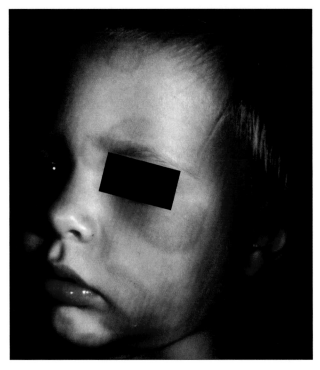

Top: Multiple rashes of erythema migrans on a child with early disseminated Lyme disease. Photograph by Alan MacDonald, previously published in Alan G. Barbour, "Laboratory Aspects of Lyme Borreliosis," *Clinical Microbiology Reviews* 1 (1988): 400.

Bottom: Next to a dime are the larva (smallest), nymph (intermediate size), and adult female (largest) of *Ixodes scapularis*, the deer tick of the central and eastern United States. Photograph courtesy of Russell Johnson.

If untreated, the ringlike rash or patch may grow during the next two to four weeks until it is more than a foot in diameter. More commonly, the rash begins to recede on its own after a certain point: about four weeks in North America and up to ten weeks in Europe. How could this occur, since the spirochetes are so adept at staying one step ahead of the advancing inflammation? The explanation lies in the fact that the person eventually launches a more formidable defense against the infection than the nonspecific first-line defense of innate immunity. This later response is a different matter. It is tailor made to combat the particular type of microorganism that is invading.

The Specific Response to Infection

Antibodies are the slower developing but more specific response to the spirochetes. Antibodies are the special-purpose proteins that are produced after we experience a new or foreign substance in our bodies. The event that introduces the new substance may be an infection—having one bout of influenza protects us against having another bout with the same strain of virus. It might also be a vaccination; after a tetanus shot, we are protected against developing that disease for several years. Antibody proteins derive their specificity from unique regions in their structures. These regions fit around a part of the influenza virus or a part of the tetanus poison as a lock fits around a key. We speak of an antibody "recognizing" this or that substance. An antibody that is effective against a tetanus poison is not effective against the influenza virus, and vice versa. This type of immunity is termed "adaptive immunity" to distinguish it from innate immunity. Humans and other vertebrates have this defense that can "adapt" to what the infectious threat is. The combination of innate immunity and adaptive immunity is usually enough for us to get over most routine infections, like the common cold virus or an infectious diarrhea, such as from *Salmonella*.

Antibodies that recognize the Lyme disease spirochete may be present in the host body before the infection occurs but not in large enough numbers to be effective in curbing the spirochetes' initial spread. It may be several days to weeks before there are sufficient numbers of antibodies to launch an adequate attack against the spirochetes. Case F was an example of this. The blood sample taken a week into the illness did not have antibodies, but the one about seven weeks later did. The average time for there to be enough antibodies to be detected by laboratory tests is about two weeks.

If antibodies are one arm of adaptive immunity, the other arm is the lymphocyte. Lymphocytes are a type of white blood cell. When a white blood cell (WBC) count of the blood is performed, it is generally found that lymphocytes make up the majority or a large minority of the white blood cells in the circulation. Some lymphocytes in the body respond in a specific or restricted way to infectious microorganisms. Indeed, one type of lymphocyte produces antibodies. Other lymphocytes do not form antibodies but either help to produce them or perform much as antibodies do, recognizing and attaching themselves to parts of bacteria or viruses. Like an antibody, a lymphocyte of this type, called a T-cell, can respond only to that molecule or part of a cell for which it is particularly fitted. The contribution of this type of lymphocyte to an effective attack or immunity against Lyme disease spirochetes appears to be less important than antibodies in halting the infection in its early stage. Indeed, T-cell lymphocytes are probably more important in causing inflammation and tissue damage, such as in the joints or in the nervous system, in Lyme disease than in protection against the spirochetes.

For some people who become infected with Lyme disease spirochetes, this combination of innate and adaptive immunity not only limits the erythema migrans rash but is sufficient to cure them of the disease. They are no longer ill from the infection. The spirochetes have been either completely eliminated from the body or so restrained that they no longer cause harm. In such cases, the only illness the infected people suffer is the rash and some of the more general symptoms that commonly accompany it, like headache, muscle aches, and tiredness. Their temperature may be one to two degrees Fahrenheit above normal at most. This set of symptoms is sometimes called flulike for its resemblance to some symptoms of influenza.

Other localized symptoms of early infection may include enlarged lymph glands or nodes near the rash. Symptoms that are rare during this stage of illness are runny nose, coughing up sputum, rash on the palms or soles, diarrhea, vomiting, and abdominal cramps. If these symptoms occur and there is no rash, the chance that the infection is Lyme disease is minimal.

For many people, the infection is limited to around the tick bite site, either from the natural immunity that develops or from antibiotic treatment. In others, however, the infection spreads widely from the original site of the bite before an effective immune response develops. Spirochetes accomplish this by first entering from their location in the skin to the blood or the lymph, which is like blood but without red cells and in separate vessels. In the lymph system, they travel toward the center of body and reach the nearest lymph gland, causing enlargement. If they get in the bloodstream, they can circulate for one to three weeks.

Invasion of the blood by the Lyme disease spirochete happens in about one-half of patients with erythema migrans in North America and less commonly in Europe.[6] Although the number of spirochetes is low—only about one bacterial cell in 10 milliliters of blood—patients with blood infection are more likely to have fever and additional symptoms, like headache, mild neck stiffness, muscle and joint aches, tiredness, and loss of appetite, than those with infection limited to the skin where the tick bit.

Other symptoms and signs for spread beyond the skin are an enlarged spleen, enlarged lymph nodes throughout the body, reddening of the whites of the eye, or swelling of the testicles. A definite sign of a disseminated Lyme disease is the presence on the skin of multiple erythema migrans rashes, most of which are distant from the site where a tick originally attached (see photo on page 20, top). Some types or strains of the spirochetes are more capable of spreading distantly in this way than other strains, which are more limited in their capacity to cause disease. Multiple erythema migrans rashes are more common in North America than in Europe or Asia.

Eventually, antibodies attach to spirochetes in the blood and remove them from the circulation. By the time that occurs, however, some spirochetes have left the blood and entered distant organs by penetrating the linings of small blood vessels. Once they reach the other side of the blood vessels, spirochetes can reside and move in the liquid between cells in other tissues and organs, including the joints, heart, and brain. By some method, the spirochetes in these other locations are effectively hidden from the roaming antibodies and lymphocytes. Whether the spirochetes move to and persist in organs and other tissues deep inside the body is one of the critical events that determines whether a person has only erythema migrans or something more serious, as the next chapter describes.

Does Asymptomatic Infection Occur?

In a human vaccine trial, the study subjects, some of whom received the vaccine while others received the placebo injection, were followed closely after their immunization to see if they got *B. burgdorferi* infection during the follow-up period.[7] The evidence of infection was the new appearance in the interval of antibodies to *B. burgdorferi*, whether they had a rash or not. Of 269 individuals who met this criterion for infection, 42 did not have erythema migrans but did have other symptoms. In 18 of these, there was evidence of another tick-borne infection besides *B. burgdorferi* (chapter 8). The other 24—about 10 percent of the total—appeared to have had a recent *B. burgdorferi* without a rash. (Why is this figure lower than the 20- to 30-percent figure often quoted for Lyme disease cases without a rash? Probably because both the participants and the study investigators were particularly assiduous in looking for rashes.)

But another 8 individuals in the study were truly asymptomatic and were not treated at that point. One of these 8 later presented with arthritis of Lyme disease. So asymptomatic infection can occur, but it appears that the large majority of people in North America who get infected become sick from it to one degree or another. In Europe, asymptomatic infection, that is, the presence of antibodies to the spirochete but no history of an illness compatible with Lyme disease, seems more frequent. As we will see in chapter 3, some types of the Lyme disease spirochetes found on the Eurasian continent are less capable of spreading beyond the tick bite site.

Can Lyme Disease Be Passed to the Fetus during Pregnancy?

In the 1980s, there were two reports of transmission through the placenta to the fetus of *B. burgdorferi* from mothers who had Lyme disease during the first trimester. Both infants died soon after birth. Autopsies found evidence of spirochete-like bodies in tissues examined by microscope. Because there was no culture of the bacteria or any tests of antibodies, the evidence of *B. burgdorferi* infection in the infants was inconclusive. In a subsequent U.S. study of nineteen cases of Lyme disease during the first, second, or third trimester of pregnancy, five cases had a poor outcome for the fetus or infant.[8] But the outcomes were different in kind; there was no common pattern, such as what occurs in congenital syphilis. So again, there was not a conclusive link to the mother's infection. In a larger study of 105 women with erythema migrans during pregnancy in Europe, 89 percent had full-term healthy infants, 6 percent had premature infants, and 2 percent had miscarriages.[9] Of the 93 full-term deliveries, 4 had congenital abnormalities, but these were not attributable to *B. burgdorferi* infection. In other words, there were other plausible explanations.

If Lyme disease during pregnancy does directly cause stillbirths and sick or abnormal babies at all, it does this rarely enough that the effect cannot be noticed in studies comparing areas with high and low risks of Lyme disease.[10] These findings should by no means lead to complacency about Lyme disease in women of child-bearing age. It's important to take extra precautions to prevent all kinds of infections in pregnant women. If Lyme disease does occur during pregnancy, treatment should be speedy and cognizant of the consequences for the fetus as well as for the woman (chapter 9).

Are People with Lower Resistance at Special Risk of Complications from Lyme Disease?

Another group of people who might be expected to do poorly with Lyme disease are those with lowered resistance to infection. If, as we have seen, the immune system can ultimately control the infection in many if not all people, then shouldn't those with defective immune systems be at a higher risk of infection and more serious disease? People with HIV infection and AIDS come to mind, as do people with transplants, cancer, or autoimmune disease who are on drugs that suppress the immune system. We could also include the very young and older people. The infant immune system is not as effective as the immune system in the older child and adult, and after a person passes the age of 70 or so, the immune system loses some of its ability to thwart infections and check cancer growth.

There have been a few reports of a more severe course of Lyme disease, including brain involvement, in association with HIV infection,[11] but not as many as one might have predicted thirty years ago, when both the HIV virus and *B. burgdorferi* were first discovered. With AIDS and most immunity-suppressing therapies, the principal deficiency is in the T-cell lymphocytes that act against an infection, and the antibodies are less severely affected. Studies of Lyme disease in animals indicate that antibodies are more important than disease-fighting cells in controlling this spirochete infection. Thus, as long as a person has the ability to make antibodies, the infection can likely be controlled, especially when an antibiotic is added. There are also much more effective therapies for HIV infection now, so that even when a cure is not achievable, the HIV-positive individual's immune system function may be close to normal.

On the other hand, very young and very old people may have deficiencies in producing antibodies. But people in these age groups have a lower risk of Lyme disease, because they are less likely to be out of doors and thus are less likely to be exposed to ticks (chapter 4).

Chapter 2
Late Infection and Its Complications

■ **Case G.** A 55-year-old woman had a headache, neck pain, and a mild fever develop in June. After three weeks, she noted weakness over her face muscles on both sides and difficulty walking. After she was admitted to the hospital, an examination showed paralysis of facial muscle nerves, leading to drooping eyelids and an inability to make a full smile. Because of unsteadiness, she was able to stand and walk only with her feet far apart. But her strength and sensory capabilities were not impaired. A lumbar puncture was performed, and the fluid that was obtained had many white cells and elevated protein, which was an indication of an infection around or in the brain. She was treated with intravenous penicillin for 10 days, and over the following month, her symptoms disappeared. She remained symptom-free for the four years of followup.[1]

■ **Case H.** A 27 year-old resident of Connecticut had a flulike illness lasting ten days. Four months later, he experienced pains that moved about between his shoulders, elbows, wrists, hips, jaw, and fingers. The pain lasted several days in any given joint. After one and a half years of this type of migrating joint pain, marked swelling and warmth of the left knee developed. The patient was started on an aspirin-like medication. This helped with the pain, but the knee remained swollen for several months thereafter. Eventually, an operation to remove the inflamed tissue of the joint was performed, and his disability lessened thereafter.[2]

■ **Case I.** A 65-year-old man was hospitalized in November because of episodes of breathlessness during mild exertion and dizziness after rising to a standing position. While vacationing in Massachusetts in July that year, he had found a tick attached to his thigh. This was followed by a fever and a red rash with clearing in the center, where the tick had been embedded. In the hospital, an electrocardiogram revealed that the patient's abnormally slow pulse rate was from a nerve block in the heart. A temporary pacemaker was inserted during treatment with an antibiotic.[3]

In these three cases, the major consequences of the *Borrelia burgdorferi* infection occurred far from the tick bite and at a distance in time. Only Case I noted a rash, which may explain why only when the infection had spread and affected other organs did Cases G and H come to medical attention. In Case G, the nervous system was stricken. For Case H, it was the mainly the joints. And in Case I, the effects of a serious heart abnormality brought the patient to the hospital.

Of note, all three case reports were published in medical journals before 1990. They serve now as examples of the "natural history" of the infection, meaning the course of a disease in the absence of an intervention, like antibiotics. During the period when a virus was thought to cause Lyme disease and therefore patients were not treated with antibiotics, the natural history of Lyme disease, over years in some instances, was there to observe. But three decades on, the links between erythema migrans and otherwise unexplained disorders of other organs are much better appreciated. Patients with Lyme disease, even in less typical versions, are more likely to be diagnosed and treated early than in the past, when watchful waiting may have been the practice. This means that many of the complications described below are less common now than a reading of older textbooks and layperson's guides would lead one to expect. As an example, the heart disorder that Case I suffered reportedly occurred in 5 to 8 percent of Lyme disease cases in the past. Now the frequency is around 1 percent, mainly because of better recognition of early Lyme disease, not only by medical professionals but by the public.

Another consideration when assessing what might be a complication of Lyme disease is the strength of evidence for a cause-and-effect relationship. Many articles in the medical literature, especially those published in the first few years after *B. burgdorferi*'s discovery, based their conclusions of an association between the spirochete and a certain disorder on what can now be seen as shaky grounds, such as measurements of antibodies by a crude laboratory test. While there continue to be new discoveries about what *B. burgdorferi* infection can do, and enigmas remain (chapter 10), there is also recognition that some conjectures about disease causation, such as that *B. burgdorferi* infection causes multiple sclerosis or amyotrophic lateral sclerosis (Lou Gehrig's disease), could not be substantiated.

Spirochete Survival Strategies

A question that all three cases raise is this: Why would the disease progress in this way in distant locations in the body if the immune response had cleared the skin and the blood of infection (chapter 1)? That is the subject of ongoing research, but the findings to date suggest two main reasons for disease progression in the face of what seems to be effective immune response. The first explanation is that the tissues where the spirochetes are most likely to cause later manifestations are less accessible to the antibodies and phagocytes that would otherwise kill the spirochetes if, say, they were swimming in the blood. These protected locations might be behind a barrier, across which the passage of antibodies and white cells is difficult. Such places are the brain, the eye, the spinal cord, and the roots of nerves as they emerge from the brain and spinal cord. Behind this protective barrier, these parts of the body normally are not exposed to microbes, so they have less need for active surveillance by the immune system. Other locations where spirochetes can sequester are not behind a barrier per se but have comparatively poor blood supply. The antibodies and phagocytes circulating in the blood cannot carry out their function if they have limited access to their quarry. Some of the supporting tissues around joints have this characteristic.

A second explanation for infection persistence despite an immune response is the evasiveness of the spirochete. The bacterial cells can turn off the production of the main target for the antibodies when spirochetes are in the skin and blood. The antibodies are still circulating, but the protein they would bind to on the surface of the spirochetes is gone.

That still leaves another protein on the surface, against which immunity can be elicited, but a small fraction of spirochetes appearing in a new generation of the bacteria as the population grows have a slight change in the composition of that protein. This is enough of a modification of its shape for the existing antibodies to have less effect on the variant spirochetes. This new population of variants expands, and with that expansion, another set of variants appears, allowing *B. burgdorferi* to remain one step ahead of the immune system. These evasive steps are usually insufficient to allow the spirochetes to survive in the blood again, but they are enough in combination with the strategy of hiding out to allow them to persist and to continue to cause disease.

Infection of the Nervous System

Case G had Lyme disease with involvement of the nervous system. Her illness started off with a flulike illness without a rash, and the infection progressed over a few weeks to paralysis of the nerves that emerge from the brain and bones of the skull and that control most facial muscles. Examination of the spinal fluid also showed evidence that the covering around the brain and spinal cord, called the meninges, was affected. In this instance, there was a blurring between the early phase of the infection and the later phase. Although the "early" and "late" dichotomy can be useful for conceptualizing the disease, it is important to stress that there may not be a clean break in the disease, with a period of well-being between well-demarcated "stages." With that caveat in mind, we can consider the manifestations that tend to occur after the early infection, be it a flulike episode or erythema migrans, has passed.

In its natural history, the Lyme disease spirochete behaves similarly to what might be considered "cousin" spirochetes, those that cause syphilis. In both Lyme disease and syphilis, the infection begins in a surface tissue: skin in the case of Lyme disease, and the penis, the vagina, the anus, or the mouth in the case of syphilis. After a period of days or a week, the spirochetes spread from this jumping-off point through the blood to the brain and other tissues, including the heart. In syphilis it may be months to years until the symptoms of nervous system involvement appear. There may be a delay in the onset of neurologic symptoms in Lyme disease, too, but it is usually measured in weeks to months, not years.

The three most common complications involving the nervous system are (1) a type of meningitis, (2) complete or partial paralysis of some nerves for the muscles of the face and eyes, and (3) inflammation and irritation of nerves near their roots at the spinal cord. The patient may have only one manifestation, a pair of them, or all three. Between 10 to 15 percent of untreated patients with Lyme disease in North America and Europe will experience at least one of these complications within a few months of the infection's start. I consider each in turn.

A headache during Lyme disease may be a consequence of the cytokines that are released systemically. It is a manifestation of the person's nonspecific response to infection and is not a sign of actual nervous system invasion. In this situation, the headache is not caused by spirochetes in or around the brain itself. However, if the headache is particularly severe, persists for more than a few days, and is accompanied by neck stiffness, the meninges lining surrounding the brain may be infected or irritated. This is a form of meningitis (inflammation of the meninges). The lining of the brain swells and stretches, stimulating the nerve endings in the meninges and producing headache pain and neck stiffness. The symptoms may be fluctuating in severity. During meningitis there are more white blood cells and increased amounts of protein in the fluid surrounding the brain and spinal cord, as we saw with Case G. Typically, during Lyme disease the white cells in the spinal fluid are mainly lymphocytes rather than the phagocytic white cells that one would see in life-threatening forms of meningitis from other types of bacteria, like the type that can cause outbreaks in schools and dormitories.

To distinguish it from the medical emergency form of meningitis, this complication of Lyme disease is referred to as a "lymphocytic meningitis." It may also be called "aseptic meningitis" by a physician, where "aseptic" refers to the failure to recover by routine culture one of the life-threatening bacteria. Aseptic meningitis can also be caused by one of many viruses and therefore is not a sure sign of Lyme disease by itself.

A common complication involving the nervous system is weakness on one side of the face, or less commonly, both sides of the face, in 5 to 10 percent of patients with Lyme disease. This can be alarming, because it can progress over a short period and may suggest that a stroke is occurring. When the facial muscles are so weakened, one's smile is lopsided, the brow is unfurrowed on that side, and the eyelids may not fully close. It can be difficult to chew food on that side, because food particles get stuck between the teeth and cheek. The speech may slur, and it is hard to blow. Paralysis of these muscles is the consequence of inflammation of one of the nerves passing through a narrow passage in the skull, in this case the nerve that connects the brain to the facial muscles. This type of nerve paralysis is called Bell's palsy. The inability to close or blink the eye, or eyes if both sides are affected, can lead to cornea damage, unless steps are taken to protect and regularly lubricate the eye. Most people with Bell's palsy from Lyme disease (or from some other cause, such as a viral infection) eventually recover function of the muscles over a period of weeks, but some are left with a mild droop on one side of the face.

Other such nerves that emerge from the brain within the skull or cranium—thus, their name "cranial nerves"—control eye movements or carry impulses to and from the ear. If one or more of these other nerves become inflamed, the result may be double vision and eyes that do not line up or hearing and balance problems. These complications are less common than Bell's palsy, and, as with Bell's palsy, there are other possible causes of these nerve disorders than Lyme disease.

Limb or trunk pain, numbness, and tingling sensations (feeling like it has "gone to sleep") is a result of inflammation of the nerves as they emerge from the spinal cord in the neck and back. This complication occurs in about 3 percent of cases of Lyme disease. Some patients have weakness in the affected arm or leg. Severe pain in an arm or leg is more common in Europe than in North America. Often the symptomatic limb is the one that had the tick bite. As was true for Bell's palsy, these abnormalities of limb and trunk nerves are not, in isolation, unique to Lyme disease. There may be other medical explanations and, thus, other potential treatments.

Direct invasion of the brain or spinal cord itself and multiplication of the bacteria there can occur, but it is unusual in patients with Lyme disease and in various animals, including mice and monkeys, with experimental infections with *B. burgdorferi*. But the prospect of any illness that affects a person's ability to think, concentrate, and remember is chilling. Severe inflammation of the brain and spinal cord in response to infection and leading to severe, sometimes permanent, neurologic deficits occasionally occurs with infection by a spirochete that occurs only in Europe and Asia. This is called "encephalomyelitis" for the inflammation (-itis) of the brain (encephalo-) and spinal cord (-myel-). This condition is unusual in North America, but it can occur and may initially be confused with multiple sclerosis. Case G's difficulty in standing and walking may have been a sign that a part of her brain controlling balance and sense of body position was affected.

In North America, a now-rare late neurologic complication of *B. burgdorferi* infection is a more subtle and poorly understood disorder of the brain that may surface several months to years after the onset of infection. In this form of the disease, inflammation of the brain is milder than the more destructive version in Europe and Asia but still detectable by laboratory tests of the spinal fluid. There may be memory impairment that is documented by formal psychological testing, a mood disorder that is like major depression, and extreme fatigue. Confirming a relationship between these symptoms and active *B. burgdorferi* infection is difficult, though, for there may be many other possible explanations. It requires not only demonstration of antibodies to the spirochete in the blood but also production of antibodies within the brain itself (chapter 7).

Infection of the Joints

After the spirochetes entered the blood of Case H, they traveled to his joints and eventually caused arthritis, first in several joints in a migrating pattern and then settling into one joint, his left knee. In North America, about 50 to 60 percent of patients who did not receive antibiotics developed arthritis. The swelling and pain in one or two large joints—commonly the knee or both knees, but also the ankle, hip, shoulder, elbow, and wrist—usually follows a period earlier in the infection during which several joints are painful but not usually so swollen. Typically the pains migrate from one joint to another, and they come and go. Later, when one or two joints alone are affected, there are weeks in which the pain is intense and other weeks when the person is comparatively pain-free and the swelling subsides. The amount of fluid in the joint increases to the point where drainage through a needle provides relief. Even without treatment, the number of patients with recurrent attacks declined by 10 to 20 percent each year. But there remained a group of patients—one in ten of those who had sore joints originally—with unremitting arthritis of one or two large joints, lasting more than a year.

A sample of fluid from around the arthritic joint shows increases in the numbers of white cells and in the amount of protein. The cytokines that foster the inflammation build up in the fluid, attracting more white cells. The layer of cells that line the joint thicken. There may be erosion of the adjacent bone and cartilage that can be seen on x-ray or MRI. An exam by microscope of a biopsy from the knee joint shows the presence of lymphocytes and other types of white cells, as well as extra growth of joint lining cells. This cell growth not only causes pain but limits joint motion. The whole picture resembles—but is not identical to—rheumatoid arthritis.

With earlier recognition of the disease and timelier treatments, the late problems with one or two joints that Case H suffered are far less common. But about 30 percent of people with Lyme disease still report the more generalized joint pains, and consequently, they are at risk of the more chronic form of arthritis at some later date. The swollen, painful knees and ankles that children of Lyme, Connecticut, experienced in the 1970s can still occur, when either the signs of early Lyme disease are ignored or overlooked, or when individuals, for religious or other reasons, eschew using pharmaceutical drugs like antibiotics.[4]

What Case H experienced early on—the soreness in several joints—has its parallel in mice infected with *B. burgdorferi* infection in laboratory experiments.[5] At first the spirochetes circulate in the blood and then invade and multiply in the joints. Their presence induces an inflamma-tory response, with the accumulation of white cells and tissue fluid. The joints of the affected mice become swollen and stiff, and the affected animal hobbles around the cage. But not all breeds of mice have arthritis to the same degree. Some breeds have severe arthritis. In others, the joints may have as many spirochetes in them, but the inflammation is less. These studies show that it is the combination of the spirochete's invasiveness and the host's response that accounts for how much arthritis will occur. What numerous experimental infections have not provided is the chronic arthritis of one or two joints that humans can have. This does not occur in laboratory mice of any breed that has been studied to date.

The Question of Autoimmunity and Lyme Arthritis

Once antibiotics were routinely instituted as treatment, around half the patients with chronic arthritis responded with a slow resolution of the arthritis. In these patients, there was often evidence that the spirochetes were still present in the joint at the time the antibiotics began. But the other half of patients showed no improvement during a three-month followup after antibiotics had stopped. In these patients, there was no evidence of a continued presence of viable spirochetes.

Whether a more chronic and disabling arthritis develops is determined in part by the person's genetic background. There are certain genes, known as MHC genes, which a person inherits from his or her parents, that may put the person at higher risk of severe, long-lasting arthritis during Lyme disease. Curiously, the same genes are also associated with an increased risk of developing rheumatoid arthritis. This is one reason many researchers think that if scientists can reach an understanding of the cause of arthritis in Lyme disease, they may discover important clues for understanding rheumatoid arthritis. So far, the studies of laboratory mice have shown that the MHC genes do not alone determine whether arthritis occurs and, if so, how severe. Other parts of the genome, places on chromosomes distant from where the MHC genes are, have important roles, too.

In rheumatoid arthritis, the person may mount an immune response against his or her own tissues and cells. This type of inappropriate, potentially damaging immune response is called autoimmunity. There is some circumstantial evidence that autoimmunity occurs in certain people with Lyme disease, notably among the small number who continue to have a swollen painful joint months after a long course of antibiotics. A few years ago, there was the theory, based on a few laboratory studies, that some antibodies directed against a protein of the spirochete also reacted with human tissues, thus engendering self-directed damage in the joint. This was plausible, because antibodies can take only a limited number of shapes, and some of them by chance may bind or fit with cells and tissues of the person or animal in whose body they were created, as well as with a part of the invading microorganism. But subsequent studies could not confirm that this laboratory observation was of importance in causing human disease.[6] (And remember, this can only be studied in *people* with Lyme disease, because mice do not get the joint problem that looks like rheumatoid arthritis.)

So, where do things stand on the autoimmunity issue? We know that in both mice and humans, genetic background has a determining role in the outcome of joint infection. In mice, this is for more generalized arthritis, and in humans, the known gene association is for the more severe, single-joint arthritis. We also know that the inflammation continues in the joints long after intact living bacteria are gone. We know that some candidates in the spirochetes that could elicit autoimmune antibodies or lymphocytes have not held up under further scrutiny. Whether the culprits could be some persisting remnants, if not whole cells, of the spirochetes is discussed in chapter 10.

Infection of the Heart

Deaths as a direct result of bona fide infection by *B. burgdorferi* are rare—perhaps fewer than twenty have been reported in the world over thirty years. This is out of the hundreds of thousands, if not millions, of human infections during this period. The scarce numbers of postmortem pathology reports in the medical literature speak of the rarity of *B. burgdorferi*–attributable mortality. But if a fatality does occur during Lyme disease, it is likely attributable to either involvement of the heart or a coinfection with another tick-borne pathogen (chapter 11).

Animals with *B. burgdorferi* infection frequently have spirochetes in their heart. Spirochetes seem to preferentially settle in the heart, as well as in the joints and nervous system, in people as well as in animals. The microorganisms enter the muscles of the heart and produce inflammation there. The white blood cells increase in numbers, and the spaces between cells swell with fluid. This inflammation of the heart's muscles can reduce the strength of the heart's contractions, and in this condition, the heart cannot sustain the same load it normally would.

Case I had the most common manifestation of heart infection, or carditis: an abnormally slow heart rate. If the inflammation is in the conduction system for the heartbeat, the impulses are slowed or blocked. The electrical message for the heartbeat, which starts in the upper part of the heart, cannot reach the lower heart, where the pumping action is strongest. The lower heart has a backup conduction system that comes into play if the impulses cannot get through to it. But the backup's intrinsic rate of beating is much lower than that of the normal control system— thirty to forty beats per minute rather than sixty to eighty. At rates this slow, the heart may not be able to pump enough blood to supply the vital organs, such as the brain. The symptoms of slowed or blocked electrical messages in the heart include fainting or a feeling of light-headedness that might be relieved by lying down. This condition may necessitate the temporary implantation of an artificial pacemaker until the heart's rhythm returns to normal, which usually takes one to six weeks.

The current frequency of carditis during Lyme disease is about 1 percent. The great majority of patients recover. Up to 2012, only four medical reports of deaths in the world were attributed to heart involvement in Lyme disease. But between 2012 and 2013, the Centers for Disease Control (CDC) investigated three sudden deaths from Lyme disease carditis in people in their twenties and thirties in the United States.[7] The deaths occurred around one month after what in retrospect was the start of infection. Lyme disease was recognized only after death, when the hearts of these young adults were examined for prospective organ transplants and found to have abnormalities. The ages of these victims correspond to the overall peak age period for carditis, 15 to 44 years. Overall, carditis cases are more common in the 15- to 44-year-old age group. Although 70 percent of Lyme disease patients have an erythema migrans rash, less than 50 percent of the carditis cases did. The lesson here is that although serious disorders of the heart during Lyme disease are uncommon, young adults may be at special risk of sudden death, perhaps because they are not as likely to seek early medical attention as older people or parents of children who become sick.

Other Complications of Lyme Disease

The nervous system, the joints, and the heart—these are the organs and tissues commonly affected by the spirochete when it spreads in the blood. But other parts of the body may be involved. The eyes are another location where the spirochetes are less accessible to antibodies and phagocytes. When one or both eyes are affected, there may be a change in vision and an increased sensitivity to light from infection or secondary inflammation within the globe.[8] Redness of the whites of the eyes is usually not by itself a sign of Lyme disease; there are many other explanations for this symptom besides Lyme disease, and usually the vision is not affected. But keratitis, an inflammation of the cornea, the front part of the eye through which light passes, occurs in some patients with Lyme disease. It can be very painful and can lead to damaged vision.

Early studies of Lyme disease patients in North America revealed minor abnormalities in their blood. These abnormalities, discovered during testing, were of some use to physicians in making the diagnosis if their suspicion of Lyme disease was already high. But the abnormalities were not specific for Lyme disease—that is, other medical conditions produced similar abnormalities in the blood. Since the discovery of the Lyme disease spirochete and the availability of a specific diagnostic test, these blood abnormalities have assumed less importance. Nevertheless, the types of abnormalities observed provided additional insight into how the spirochete affects people.

What physicians found was evidence of mild liver damage and increased production of all antibodies, not just the ones against the spirochete. The liver damage discovered in a minority of patients was detectable only by blood tests—physicians discovered no evidence of a problem with the liver simply by talking to or examining the patient. This situation, in which an abnormality in a blood test of liver function is present without evidence of significant disease of the liver, occurs frequently in medicine. It probably results from the liver's tendency to release its enzymes and other proteins into the bloodstream after relatively minor insults and stresses. In any case, finding higher levels than normal of liver enzymes in the blood is not uncommon during Lyme disease and does not usually indicate a serious problem with that organ. Only a few patients with Lyme disease have had liver inflammation to the point of developing true hepatitis.

The broad increases in antibodies during active infection are not peculiar to Lyme disease. Other infections and inflammatory states, like rheumatoid arthritis, may also turn up the dial for more production of antibodies of all sorts. This increase usually has little consequence, except that some Lyme disease patients may have borderline or even positive reactions against other microbes in laboratory tests. These may include microbes from other tick-borne diseases, such as the *Babesia* parasite, covered in chapter 11. So a positive result in a test for antibodies to the *Babesia* may be a false-positive if performed during active Lyme disease. A repeat *Babesia* test after the *B. burgdorferi* infection has resolved may be negative.

What Parts of the Body Aren't Affected by Lyme Disease?

After this litany of troubles caused by the Lyme disease spirochete, it may seem that no part of the body remains uninvolved when this infection occurs—but that's not so. Most of the body's organs and tissues—for instance, the lungs, the kidneys, the bones, the pancreas, the gallbladder, and the intestinal tract from the esophagus to the anus—are rarely if ever affected in Lyme disease. Most of these places would be inhospitable to the spirochetes. This includes the bladder, which is frequently infected in rodents but apparently not in humans. If there is evidence of infection or other abnormality in these other body locations in humans, it indicates that the disorder is either not Lyme disease or includes something else in addition to Lyme disease.

In Lyme disease, too, there is seldom any abnormality of the glands that produce hormones, such as the thyroid and adrenal glands. Lyme disease by itself does not appear to increase the risk of a heart attack from blockage of an artery. It does not cause cancer, diabetes, hemorrhoids, periodontitis, leukemia, osteoporosis, psoriasis, heartburn, ulcerative colitis, migraine headaches, or a hernia. While the bacterium *Helicobacter pylori* is now known to cause stomach and duodenal ulcers and stomach cancer, *B. burgdorferi* cannot make that claim. There is no evidence that it causes a condition like AIDS or other forms of serious depression of the immune system. Despite the claim by some that *B. burgdorferi* infection can cause autism, there is no convincing evidence of that. It does not cause multiple sclerosis, amyotrophic lateral sclerosis (Lou Gehrig's disease), or Alzheimer disease, though patients with these and the other conditions listed can also have Lyme disease or have had it in the past.

What Happens When Dogs and Other Domestic Animals Get Lyme Disease?

The ticks that carry the Lyme disease spirochetes bite pets as well their owners. Among all the domestic animals at risk of infection, dogs appear to contract the disease most often. In some areas of the United States, the majority of dogs in the community appear to have been exposed to the Lyme disease spirochete.[9] Most dogs do not get sick from the infection, but some develop a lameness in the legs that is equivalent to the arthritis of Lyme disease. This is more common in puppies and younger dogs. There is pain, warmth, and swelling of one or more of the affected joints. The dog refuses to bear weight on the limb with the sore joint. The lameness may be accompanied by fever, lethargy, listlessness, and loss of appetite, and the arthritis may come and go over several months. As with people, the onset of the arthritis may not occur until weeks to months after the dog's exposure to ticks. A few infected dogs also develop disease of their kidneys, a complication that has not been observed in humans and that can be fatal.

All breeds of dog appear to be susceptible to infection with the spirochete. Indeed, dogs generally are at a higher risk of infection than are people. Dogs, especially those allowed to run free, have a higher exposure to ticks than most people do. In contrast to humans, who usually acquire the infection from nymphs, a juvenile stage, dogs are more commonly infected by an adult tick than by a nymph, probably because the smaller nymph has a more difficult time than an adult tick penetrating the thick fur of the dog. Adult ticks' preferred hosts are larger furred mammals, especially deer.

There is evidence that cats, cows, horses, and sheep can be infected with the Lyme disease spirochete, but the information on infections in these other animals is not extensive. The most compelling data exist on infections of horses. There are reports of lameness, with tender, swollen joints, in some horses living in areas where Lyme disease is common.

Chapter 3

The Pathogen, Its Vector,
and Its Reservoirs

The Pathogen: Spirochete Biology

Borrelia burgdorferi and untold other bacterial species occupy a kingdom separate from other forms of life. The group of bacteria that contains the Lyme disease agent is the subdivision known as spirochetes. Spirochetes are one of the most distinctive forms of bacteria, in their appearance and in their behavior. They have a spiral or wavy shape (hence their name), and they move through liquids and the tight spaces between cells, either as a screw moves through wood or as a snake moves over the ground. The movement is accomplished through the action of a biological motor that rotates or undulates the spirochete. Like other bacteria, spirochetes are not visible to the unaided eye. A microscope that magnifies at least a hundred times is needed to see most spirochetes. Even then, the spirochete's extremely narrow width requires special optics* that highlight the bacteria to make them appear wider than they actually are.

Spirochetes have two membranes—outer and inner. The inner membrane encloses the vital materials of all living cells: DNA for the genome, RNA molecules for messengering, and proteins for enzymes and important cellular structures, all in a liquid with salts, sugars, fats, and other dissolved proteins. Located between the two membranes is the cell wall, which provides the skeleton for the cell's elongated shape. The cell walls are composed of multiple layers of long strings of sugars, which in turn

---

* Dark-field microscopy or phase-contrast microscopy are used to visualize spirochetes.

are cross-linked by a few amino acids in a row. Stripped of all attachments, the cell wall looks like a thick hairnet without an opening. Bacterial cell walls are not found in other types of life; animals, plants, and fungi don't have these structures. A drug targeting the cell wall has less effect on our own cells than a drug operating on a structure or pathway that animals and bacteria have in common.

The outer membrane of Lyme disease spirochetes is like a soap bubble, but not nearly so fragile. It also usually takes on the spiral shape of the rest of the cell, unless under very stressful conditions, in which case it may blow up to spherical proportions.[1] Some have referred to these more spherical forms as "cysts,"[2] but they are not like the more robust cysts of some parasitic protozoa and worms. And there is no convincing evidence that the "cysts" help the spirochetes make it through a rough patch, like exposure to antibiotics. What the outer membrane does have in common with a soap bubble is the fluidity of the surface. A protein that is embedded in the outer membrane is not anchored in place; it may passively move through the membrane to another location. Some of these proteins form into doughnut-like complexes, with holes that pass through the membrane and allow entry of nutrients (and antibiotics) and exit of other molecules. Other proteins are attached via stalklike extensions to the outer membrane and are peppered over the surface. These proteins interact with the cells and tissues of their hosts and tend to be different depending on whether the host is a tick or a vertebrate. Outer membranes can also pinch off and be released as even smaller bubbles or vesicles. When a lot of these vesicles are in the tissues or blood, they may amplify the effect of the spirochetes on the innate immune system and elicit more inflammation than would be expected for the number of whole cells present.

The Pathogen: Spirochete Types

The spirochete group is further subdivided into many and diverse varieties. The types differ from one another in size, physiology, and the environments they thrive in. A few of the ecologic niches individual species occupy are marsh mud, thermal hot springs, clam intestines, and the crevices around our teeth. Spirochetes in mud and hot springs are known as "free living," which means they can live outside an animal. Others, like the symbiotic spirochetes of termites, exist and propagate only within another living being. Their hosts supply not only nutrition but also safety from climate, microscopic predators, and other hazards of the outside world. Free-living microorganisms have various ways of coping with changing and inhospitable conditions they encounter, whereas host-dependent organisms are poorly equipped for ventures beyond the animal's boundaries. *B. burgdorferi* holds its own and then some inside some mammals and birds but fares badly outside its usual hosts. It does not survive drying, exposure to temperatures much above the human body's, or being immersed in pure water or highly salty water. Spirochetes do not have a thick capsule and are susceptible to detergents and soaps, which can dissolve their membranes. Unlike some bacteria, such those that cause anthrax and tetanus, spirochetes do not make hardy spores that allow them to endure for years in the soil or on rusty nails. The upshot is that one cannot get Lyme disease from shaking hands with someone, consuming food or water, toilet seats, or being nearby when someone else coughs or sneezes.

Some host-dependent spirochetes, including *B. burgdorferi*, can be grown in the laboratory outside an animal, but special conditions are required. For example, the culture medium for growing Lyme disease spirochetes contains about fifty different ingredients, including complex mixtures of proteins, fats, and vitamins. Some of these nutrients are provided by added serum from a mammal, such as a rabbit. The medium also contains a little gelatin, which thickens it to make it more like the viscous fluids in tissues. Glucose is the simple sugar that spirochetes mainly use for energy. For best growth, these spirochetes need a little oxygen, but not too much. In a test tube filled with culture medium, *B. burgdorferi* prefers the bottom half of the tube, away from the higher oxygen concentrations at the top.

Another type of spirochete causes syphilis and is in a different genus from the *Borrelia*. The syphilis agent, *Treponema pallidum*, infects only human beings, so *Homo sapiens* is the "reservoir." Syphilis is spread from person to person by sexual and other forms of intimate physical contact.* Unlike *B. burgdorferi*, with its tick "vector" used for moving from one mammal to another, there is no arthropod vector of syphilis. Unlike *B. burgdorferi*, *T. pallidum* has not been cultivated in the laboratory. The spirochetes that cause syphilis have the ability to penetrate into the brains of those infected and to remain there for extended periods, causing temporary or permanent brain damage. One of the most feared results of syphilis in the days before antibiotics was infection of the brain, and many patients with chronic syphilis of the brain populated the mental hospitals of the eighteenth and nineteenth centuries.

As its first name tells us, the Lyme disease agent is a member of the genus *Borrelia*. The genus was named for Amédée Borrel, who was a microbiologist at the Pasteur Institute in Paris in the first years of the twentieth century. All species of *Borrelia* have in common the need for both an arthropod vector for transmission and a mammal, bird, or reptile host for a reservoir. Before the discovery of the cause of Lyme disease, *Borrelia* species were primarily known as the agents of relapsing fever (chapter 11). During relapsing fever, the patient suffers with recurring episodes of high fever, each lasting two or three days, but without a prominent rash or arthritis. With the exception of one species that is transmitted by the human body louse, all of the several other *Borrelia* species that cause relapsing fever are transmitted by ticks. Different species of *Borrelia* are found on the continents of Africa, Eurasia, North America, and South America. In North America, the tick-borne forms of relapsing fever occur in western states and provinces and in some regions of Mexico as well as Central America. Most of the relapsing fever species can be grown in the same culture medium that is used for *B. burgdorferi*, and many can be studied in experimentally infected rodents in the laboratory. A newly discovered relapsing fever species is *Borrelia miyamotoi*, which is transmitted by the same ticks that carry *B. burgdorferi* (chapter 11).

---

* Although there are claims on some websites and social media of sexual transmission of *B. burgdorferi*, this has never been proved and is highly unlikely to occur in humans.

The Pathogen: Different Species Can Cause Lyme Disease

Our attention up to now has been on one species, *B. burgdorferi*. That's appropriate. It was the first to be associated with Lyme disease and is the only one known to cause Lyme disease in North America. But in Europe, it is only third on the list of most frequent causes of Lyme disease. In Asia, *B. burgdorferi* is rare if it occurs at all. On the Eurasian continent, the two most common species that cause Lyme disease are *Borrelia afzelii* and *Borrelia garinii*.* Both have species names honoring early chroniclers of erythema migrans (Afzelius in Sweden) and the accompanying nerve disorder (Garin in France). The three Lyme disease species have more in common with each other than they do with relapsing fever *Borrelia* species. Yet, there are nontrivial differences between *B. burgdorferi, B. afzelii,* and *B. garinii*. These include greater tendencies for *B. burgdorferi* to cause arthritis, for *B. afzelii* to cause, in its late stages, a skin disorder called acrodermatitis chronica atrophicans, and for *B. garinii* to invade the central nervous system. Telling them apart by a comparison of the sequences of one or more of their genes is straightforward and routinely performed. On the other hand, the three species have about the same susceptibilities to different antibiotics, so treatment options are similar in North America as in Europe and Asia. The three species also have enough in common in their sets of proteins for antibodies elicited by infection with one species to bind to some of the proteins of another species. The latter phenomenon has relevance for interpretation of the results of diagnostic tests (chapter 6).

The Pathogen: Different Strains of *B. burgdorferi*

Look up the English word "species" in a dictionary and marvel at the disparate definitions, present and past. But its meaning in biological con-

---

* Some other species in Eurasia, like *Borrelia lusitaniae* and *Borrelia spielmanii*, have been uncommonly associated with human disease.

text is precise and restricted: "a group or class of animals or plants (usually constituting a subdivision of a genus) having certain common and permanent characteristics which clearly distinguish it from other groups."[3] The official acceptance of a new species name follows a formal process in which the evidence is assessed and judged by experts.

In contrast, the meaning of the term "strain" is harder to pin down; there is usually no formal review of a new strain name. Substitute names exist for "strain." Referring to a bacterium, one might speak of it instead as a particular "type," "line," or "variant" and not get into trouble. For purposes here, a strain of a *Borrelia* species is analogous to a breed of a dog: same species—*Canis familiaris*—but with clearly distinguishing features that will be replicated in the next generation if purebred. One could also make an analogy to cultivars of roses.

The physical differences between strains of *B. burgdorferi* (or *B. afzelii* or *B. garinii*) are less obvious than, say, between a toy poodle and a Great Dane, but no less distinctive to the discriminating observer. The most evident distinguishing features between strains are in a handful of proteins embedded in the outer membrane and exposed at the surface of the spirochetes. These proteins are what the bacterium uses to contact the outside world. One of these surface proteins, called OspC (outer surface protein C), sufficiently differs between strains that an antibody that kills the OspC of one strain is not effective against the OspC protein of another. In any geographic area where Lyme disease *Borrelia* occurs, between ten and fifteen strains are present, with some more common than others. Why not just one? Or why not one hundred? It's the outcome of balancing competition between strains.

The Pathogen: Strategies for Success

Compared to the super-bug bacteria that cause plague or anthrax, which can lead to death in a few days, the Lyme disease *Borrelia* species are mild mannered. Mutant laboratory mice that have no capacity for making antibodies can live out their usual life spans with persistent *B. burgdorferi* infections. They suffer arthritis but not the overwhelming infections that would occur with most other bacterial pathogens under such circumstances. The numbers of spirochetes do increase in the infected animal but only to levels that would optimize the chances of being passed on to a biting tick. There seems to be a limit to their growth potential, even in animals with immune deficiencies. This limiting factor may be the force of innate immunity—those more primitive responses we make to invading microbes before adaptive immunity kicks in—or some self-regulator of population size in the spirochetes. More likely, it's a combination of both.

At first hearing, this low-intensity virulence sounds like a weakness for a would-be pathogen. Are these *Borrelia* species on their way to becoming more like the plague bacterium, infection with which turns out badly for the host? Probably not. In fact, it's more plausible that they are on an evolutionary path toward even less virulence, as the section below on reservoirs discusses. Yet, this is not to say that *B. burgdorferi* and those other Lyme disease species do not have their devices to succeed in both their arthropod vector and their vertebrate reservoirs. The facts are that they do gain a foothold, disseminate, and proliferate, all the while countering or evading first innate immunity and then adaptive immunity. Their success is measured not in killing the host or making it sick but in being transmitted to a new host—to live to fight another day.

What are those devices? For those with this particular interest, an account could fill chapters.[4] But here's a sampling. To achieve the first step of gaining a foothold in a new host, the spirochetes have proteins on the surface that bind them to skin tissues. To move out from the tick bite site, the spirochetes traverse the spaces between cells with their snake-like movements and resist the efforts of white cells to gobble them up. To disseminate in the blood, the spirochetes have other proteins on the surface that protect them from host molecules on alert to attack bacteria on first contact. To persist in the host in the face of rising specific antibodies, the spirochetes continually change one of their proteins on the surface to a form that the antibodies du jour do not recognize. They thus stay one step ahead of the immune system. By this time, there may be enough spirochetes in the skin for the odds to favor transmission to a tick. For that transition, the spirochetes turn off some proteins that aided the vertebrate infection and turn on others that better suit them for life in an arthropod, a very different environment. All this switching on and off of proteins is controlled by a fine-tuned mechanism that senses whether the spirochete is in a tick or in a vertebrate and then accordingly regulates which genes are active or not. There are other signals when a tick begins a blood meal that start the spirochetes on their way from the intestine into the salivary glands of the tick and then out again into a new vertebrate host. And the cycle repeats.

The Vector: An "Outside" Parasite of Animals

A tick bite hurts less than a bee sting. Indeed, most tick bites are not even felt by the person who is being bitten. And ticks themselves are not as irritating as the mosquitoes or biting flies that swarm around one's head. Yet for most people, ticks seem to be more repulsive and chilling than other blood-sucking or stinging arthropods. Ticks are in the same disgust league as leeches and tapeworms. Ticks and leeches are parasites like tapeworms, but on the outside rather than the inside of the host's body.

For all the blood they take, mosquitoes spend little time on a person's body; they fill up within minutes through their tubelike mouth and move on. The types of ticks that transmit Lyme disease are more leisurely feeders, preferring a long meal with their heads burrowed into the skin. It's very unpleasant to discover a partially embedded tick, purplish and swollen with blood, on one's own body or the body of a loved one. Fastened tight like that, a tick cannot simply be brushed away like a gnat or a flea. Removing the creature takes skill and patience, if the mouth and head are not to be left behind in the skin (chapter 13). Finding a tick crawling on clothes or skin is better than finding it after it has become embedded; on the other hand, finding one tick means that another one may have escaped detection and settled in elsewhere on the body for a feed.

But ticks are more than creepy nuisances. They are ideally suited for passing on infectious microorganisms. Bacteria, viruses, and protozoan parasites that populate the blood of their hosts may be carried from one host to another by a tick. The tick that parasitizes people and other animals is itself taken advantage of by its smaller passengers. Because the interval between feedings for a tick is measured in months and even years, a microorganism that hitches a ride in the tick until reaching its next vertebrate host has to be adapted for life in the different environment within the tick. Some tick-borne infectious agents spend more time in the arthropod than they do in a mammal or bird.

The Vector: Ticks Are Arachnids

Ticks are members of the phylum Arthropoda, the most diverse and populous group of animals on Earth. The arthropods include arachnids, insects, centipedes, millipedes, and crustacea, such as lobsters and crabs. Like other arthropods, ticks have their skeletons, or supporting hard body parts, on the outside instead of inside. Most arthropods have a body divided into separate segments. Their internal temperatures change as that of the environment changes, instead of remaining at the even warm temperature of mammals and birds. That's one reason ticks are not active and exposed during the winter in areas, like the northeastern United States, where temperatures regularly drop below freezing and humidity may be low. (But they are a threat to people during the late fall, winter, and early spring in coastal California and in the southeastern states, with the milder, moister climates in those regions during those seasons.)

Arthropods do not have true blood or blood vessels. Instead, their organs are bathed directly in a liquid carrying oxygen and nutrients. Arthropods cannot make antibodies; they do not have lymphocytes, the adaptable and protective cells of our own immune systems. Instead, they contend with infectious agents with a few preformed chemicals and cells that serve as scavengers. Compared with a vertebrate's immune system, an arthropod's is primitive and limited. Hitchhiking pathogens need only contend with innate immunity in their arthropod vectors.

Among arthropods, ticks are in the group Arachnida, which also includes spiders, scorpions, and mites. Ticks and mites are closely related and form a group, Acari, separate from other arachnids. Some tiny mites are common, harmless residents of people's eyelash and hair follicles. Another type of mite causes the skin disease scabies. Other mites are the carriers of serious diseases, like infections similar to Rocky Mountain spotted fever. A type of mite that parasitizes bees has been implicated as a cofactor in the colony collapse disorder that has decimated honeybee populations in many parts of the world.

Adult arachnids have eight legs, or appendages, instead of the six characteristic of insects. Ticks and mites differ from other arachnids in having an unsegmented body. Arachnids have simple eyes and tactile body hairs to feel their environment. Some can sense the presence in the air of certain chemicals, such as carbon dioxide, and odors an animal gives off, and this ability guides them to an animal. They achieve this guided-missile-like behavior with a simple nervous system that senses features of the environment and directs their motor parts to respond accordingly. But this is also a vulnerability. Tick nervous systems make good targets for pesticides and repellents (chapters 12 and 13).

The Vector: Life Stages

Like other types of arthropods, ticks go through different stages between hatching from an egg and achieving adulthood. Growth occurs by molting. The outer shell, or exoskeleton, is shed, revealing what had been under development below. For most tick species, including the Lyme disease vectors, as they pass through the various stages between molts, they not only become larger but change in appearance. The earlier forms are not just smaller versions of the adult. (See the color photograph of the tick's life stages in chapter 1.)

The first stage after hatching is the larval. The larval tick has three pairs of legs instead of the four pairs of later stages. The untrained eye might confuse it with an insect at this stage. Once a larval tick has a blood meal, it molts to the next stage, the nymphal. The nymph is bigger than the larva and has its own unique features. For a nymph to change into an adult, the final stage, it also must have a blood meal. Male and female nymphs look the same, so they cannot be told apart except by a DNA test. With the molt from nymph to adult, differences between sexes become readily apparent in most tick species. The generally smaller male adult need not feed; its overriding program is to find a female tick on the same host vertebrate. The adult female must feed if it is to have the energy, nutrients, and body mass necessary to produce eggs.

For many types of ticks, the interval between the different stages is usually a year, sometimes two years. A larva may change to a nymph after its meal in the spring, summer, or fall, depending on the location and climate. The nymphal form of that tick has to survive the winter before it feeds. For ticks that are exposed in the environment and depend on a passing vertebrate for their next meal, the percentage at each stage that succeed to the next stage or produce a clutch of eggs is generally low. Most never reach the finish line.

The Vector: The Blood Meal

Like vampires, ticks take their sustenance from blood. Though they don't have Dracula's fangs, ticks do have other types of specialized parts. The tick's mouth is adapted for piercing and sucking. Its teeth are turned backward to hold fast in the skin. It secretes something like a cement to keep it in place. Generally a tick's meals are far apart—separated by months or years—so it must make good use of its opportunities. For a tick to obtain a year's worth of blood, it must remain unnoticed for several hours to days. The bite provokes no pain, and the saliva that the tick injects into the skin contains many substances that prevent the clotting of blood and inflammation of the surrounding area. If the buried head and mouth of the tick were to produce itching and pain, the person's attention would be drawn to the feeding tick—and the person would remove it.

The tick's structure allows its abdomen to swell with blood and tissue fluid to many times its usual size. In this bloated state, ticks are more easily noticed on the body, but by then it may be too late to stop the infection. Once the blood enters the tick's intestine, it is slowly absorbed into cells lining the gut, where it is broken down to its constituent proteins, carbohydrates, nucleic acids, and fats, the simple chemicals that the tick's own cells can use. To make room for more blood, the tick concentrates what it has consumed by excreting excess liquid out through the saliva. The feeding tick sends water, salts, and its own waste products, but not the blood cells, back into its host. Over a few days of feeding, a single tick can take in a tablespoon (15 milliliters) or so of blood, retaining only a small fraction of that volume. For grazing animals, like cattle or zebras, repeated heavy infestations of ticks produce significant anemia.

If the tick feeds on an animal infected with a Lyme disease *Borrelia* species, some spirochetes may enter the tick's intestine along with the blood meal. They will survive and may proliferate inside the tick's gut, an environment hospitable to them. The inside of a mosquito or biting fly is not conducive to spirochetes' survival and growth because, unlike ticks, mosquitoes and flies digest blood inside the tube space of the intestine itself instead of within the cells in the lining of the intestine. The chemicals and enzymes that break down the blood in the mosquito's intestine also kill the spirochetes. This is one reason that mosquitoes and biting flies do not transmit Lyme disease.

In unfed ticks, spirochetes from the last meal remain in the intestine. Once feeding begins again, spirochetes pass through the intestinal cell layer into the body cavity itself. From there, they may migrate to the tick's salivary glands on either side of the mouth parts in the head. From this launch point, the spirochetes enter a new vertebrate host to start infection anew. This journey takes about 36 hours.

Lyme Disease Ticks

It is as simple as this: no ticks, no Lyme disease. Only ticks of the genus *Ixodes* are capable of spreading this infection. Other ticks, with genus names such as *Amblyomma, Dermacentor, Haemaphysalis, Ornithodoros,* and *Rhipicephalus,* bring their own perils in various parts of the world, but not Lyme disease.

In Europe early in the twentieth century, the skin rash erythema migrans was observed to follow the bite of a particular type of tick, one common name of which was sheep tick. The full scientific name of this tick is *Ixodes ricinus.* In the 1970s, the medical detective work described in the introduction implicated a related tick, *Ixodes scapularis*, as the vector of Lyme disease in Connecticut and adjacent states.[5] In textbooks, the alternative to the scientific name is black-legged tick, but the more common usage is deer tick.

I. scapularis is also found in southeastern and south-central states. In fact, when the "father of American entomology" Thomas Say first identified *I. scapularis* as a unique species in 1821, he was studying ticks that were collected in the South.[6] At the time the tick was not conspicuous in the Northeast and was only later recorded in that and other regions (chapter 4). This northern version of the black-legged tick was for a time given its own species name, *Ixodes dammini*, and some still use that name. But a separate name was never widely accepted, and *I. scapularis* was reinstated for the northern form. Nevertheless, there is justification for discriminating the southern and northern forms, and the *dammini* species name may get an official nod again. This is not just an academic exercise. Importantly, the southern form of the species poses less of a threat than the northern form. Few of the southern ticks are infected with *B. burgdorferi*, and the southern ticks are less apt to frequent places where humans tread. They also seem more reluctant to feed on humans when given the chance. When people are bitten by southern form of *I. scapularis*, more commonly it is by an adult than by a nymphal tick.

Other *Ixodes* species are the vectors of Lyme disease elsewhere. In far western North America, from central California up into British Columbia, *Ixodes pacificus*, the western black-legged tick, transmits *B. burgdorferi* among its reservoirs and to people. *I. pacificus* also commonly carries another *Borrelia* species, *B. bissettii*, that has not been implicated as a human pathogen. *I. pacificus* ticks, like the southern form of *I. scapularis*, feed on lizards, which are uncommon or nonexistent in colder climates for the northern form of *I. scapularis*. One reason to envy a lizard is their resistance to *B. burgdorferi* infection. Some substances in lizard blood kill the spirochetes on contact, not only in the host lizard but in the intestine of the feeding tick.[7] The consequence is that lizard blood can actually sterilize a feeding tick that had become infected with *B. burgdorferi* as a larva. In the Northeast, adult ticks have a higher frequency of infection than nymphs; it is the opposite in the Far West.

On the Eurasian continent, two *Ixodes* species with overlapping ranges are the major vectors of Lyme disease. Besides *I. ricinus*, the sheep tick, in Western and Eastern Europe, *Ixodes persulcatus*, the taiga tick,* in Eastern Europe, Russia, and northern Asia is another vector of *B. afzelii* and *B. garinii*. *B. garinii* has also been found in *Ixodes uriae* ticks among nesting sea birds, like gulls and albatrosses, near the Arctic and Antarctic circles.[8] These bird ticks are not associated with Lyme disease in people, because they rarely if ever get a chance to feed on a human, let alone see one.

In most areas where Lyme disease vectors exist, there is at least one other species of *Ixodes*. These not only are rare biters of humans but are infrequently infected with a Lyme disease *Borrelia*. Yet they may have role in maintaining the pathogen in an environment by serving as an alternative vector and, thereby, a bridge between different species of mammals or birds. In other words, some of these other *Ixodes* species are not as good as *I. scapularis* in transmitting *B. burgdorferi*, but by virtue of their preference for different types of vertebrates as hosts, they can extend the reach of *B. burgdorferi* to other potential reservoirs. Pathogens whose hosts are diverse have better odds of surviving if a population crash in one of the reservoir hosts occurs. These other species, such as *Ixodes cookei* in North America and *Ixodes ovatus* in Asia, may also be vectors of other important infectious diseases of wildlife and domestic animals.

Reservoirs: Small Mammals

Ticks can be a dead end for Lyme disease bacteria. The spirochetes remain alive and even multiply inside the tick's body, but unlike *Borrelia* species that cause relapsing fever, they are not passed from a female tick to its offspring in eggs. For the spirochetes to spread in nature, for their numbers to increase, the infected tick must feed on another animal, thereby passing on the organism through that animal to other ticks—and on

---

*Taiga is a Russian word for a swampy coniferous forest.

and on—to enlarge the population. Modern-day humans are excluded from the list of potential reservoirs. On the rare chance that someone with active Lyme disease is bitten by a larval tick, and that tick successfully embedded and fed, it is more likely to fall off indoors, on the street, or in a car than in the tick's native woods. But many other types of mammals and birds are capable not only of being infected with *B. burgdorferi* but also of passing the microorganisms on to another tick, thus completing a vector-reservoir-vector cycle.

Larvae, the toddlers of tick development, feed for about four days on their first meals. The larvae acquire the bacteria by feeding on an infected host. In the case of *I. scapularis* in the northeastern and north-central United States, this is usually the white-footed mouse, *Peromyscus leucopus* (leuco- and -pus are derived from Greek words for "white" and "foot"). Although "mouse" is part of its name, it is in truth more closely related to hamsters and voles than it is to our usual idea of a mouse: the house mouse that may inhabit the woodshed or attic.*

P. leucopus is numerous in the woods and bushy landscapes of rural and suburban areas in central and eastern North America. In some regions, almost all the mice are infected with Lyme disease bacteria,[9] and in these mice, the infection persists for months—effectively the rest of their lives—providing a constant supply of the spirochetes to be taken in by ticks. The white-footed mouse develops immunity to the infecting *B. burgdorferi*.[10] But this is only enough to clear the spirochetes from the blood, not from the skin tissue, where a tick can still pick up the spirochete as it feeds. The immunity is also limited to that particular strain of *B. burgdorferi*, meaning that the mouse can be infected with a different strain, thus extending the period when the blood, as well as the skin tissue, of the animal is infectious for ticks. Despite nearly continuous infection, the white-footed mouse suffers little if any disability and shows little overt disease. This contrasts with the laboratory mouse, which commonly has arthritis.

---

*The house mouse, *Mus musculus*, is also the most commonly used animal in laboratory research. The differences between *P. leucopus* and the house mouse mean that some lessons learned in the laboratory may not be applicable to the natural reservoirs.

Other animals besides white-footed mice contribute to infection of *I. scapularis* with *B. burgdorferi* in North America. These alternative reservoirs individually may not contribute as much as *P. leucopus* does, but cumulatively they may account for half or more of the infected nymphal ticks in some areas. Key alternative rodent reservoirs in the northeastern states are chipmunks and voles. While squirrels, raccoons, skunks, and opossums may occasionally become infected, these larger mammals are less numerous than the smaller ones, so their overall impact is low. Shrews, tiny insect-eating nocturnal mammals, may be as important as white-footed mice as reservoirs in some areas.[11] But shrews are difficult to capture and to maintain in captivity, so little is known about infection in them. In the Northeast, if *P. leucopus* is absent, as occurs on Monhegan Island off of Maine, rats may serve as the primary reservoir.[12]

The white-footed mouse's range does not extend to far western North America, so that species cannot be the reservoir there. Other *Peromyscus* species are in this area,* but the western gray squirrel instead appears to be the key reservoir in parts of California.[13] Europe and Asia have no *Peromyscus* species at all. Their roles as hosts for the tick and reservoirs for the *Borrelia* are taken by voles, other types of field mice, and rabbits. European hedgehogs, those cute little spiny creatures, are sometimes infected with Lyme disease *Borrelia,* but the type of *Ixodes* tick that usually feeds on this animal seldom bites people.

Reservoirs: Birds

It makes sense that small rodents scurrying along the forest floor would pick up ticks. They brush against blades of grass and leaves of bushes where the *Ixodes* ticks are poised to grab hold. Capture an adult white-footed mouse in Connecticut and you will probably find larval and nymphal ticks attached around its ears and eyes. More surprising is the fact that birds are also hosts for ticks. Not only that, birds can also be infected

---

*One of the most common is *P. maniculatus*, misleadingly known as the deer mouse, not because of an association with deer but because of its jumping ability.

with some Lyme disease *Borrelia*, so they are reservoirs as well. Because birds tend to travel much longer distances than small mammals, they can carry infected ticks to areas where *B. burgdorferi* has not established itself yet.

But the majority of bird species have no role in the ecology of Lyme disease, either as reservoirs or as long-distance transporters of ticks. Either the bird has migrated out of the area when ticks would be biting or it rarely or never encounters ticks. The birds at most risk for tick infestation are those that search on the ground or in low bushes for food in the form of seeds or insects and other invertebrates, like worms. The American robin and other thrushes are typical of ground-foraging birds that are susceptible to ticks and, as a consequence, *B. burgdorferi* infection. In Europe, several songbirds have similar feeding behavior. In the United Kingdom, there are also pheasants, which are larger and spend much of their time on the ground.

Reservoirs: Large Mammals

Are deer reservoirs of Lyme disease? One might think so, judging by the vector's vernacular name of deer tick. Not surprisingly, then, a common misconception is that infected deer are the main sources of the trouble for humans. As we will see in the next chapter, deer are implicated in the emergence of Lyme disease, but not as reservoirs for the pathogen. Although deer are frequently bitten by infected adult ticks, they seldom develop systemic infections or pass their infections on to larval ticks. Their major importance in the life cycle of ticks is as the definitive host for adult ticks.

In Old World Europe and Asia as in the New World, deer are common wildlife animals. But unlike the parts of North America where *I. scapularis* is prevalent, large domestic animals, namely cows and sheep, frequently graze in the same areas where there are deer. These agricultural animals are other sources of blood meals for adult *Ixodes* ticks. But like deer, cows and sheep seldom become sufficiently infected to transmit the pathogen and seldom are the hosts for larval ticks. Thus, they too are more important as blood sources for the adult ticks than as reservoirs for the infection.

Dogs and cats are smaller than deer and cows but much bigger than field mice or chipmunks. In most of the countries where Lyme disease occurs, dogs and cats are companion animals and usually in the house for some of the day. They don't have pastoral or similar outdoor duties that would make them suitable as reservoirs. If they do get infected, most dogs and cats, like humans, are dead-end hosts for Lyme disease *Borrelia*. Nevertheless, they are not off the hook as accessories to an infection event. The family pet that comes into the house after roaming in the nearby forest grove or overgrown backyard may be bringing in some infected nymphs. If unattached nymphs drop off the dog or cat on a couch or bed, the people in the house are fair game.

Chapter 4

The Ecology of Lyme Disease

Nantucket! Take out your map and look at it. See what a real corner of the world it occupies; how it stands there, away off shore, more lonely than the Eddystone lighthouse. Look at it—a mere hillock, and elbow of sand; all beach, without a background. There is more sand there than you would use in twenty years as a substitute for blotting paper. Some gamesome wights* will tell you that they have to plant weeds there, they don't grow naturally.

Herman Melville, *Moby Dick; or The Whale* (1851), chapter 14

The June 10, 1922, issue of the *Inquirer and Mirror* newspaper of Nantucket Island, Massachusetts, ran a one-column story about fishermen on the sloop *Antonia* hauling on board a deer that was swimming a "long distance from shore." The fishermen brought the deer to their home port on the island and notified game warden William Jones, who took charge of the exhausted deer. After it had recovered, he released it among some pine trees out of town in the middle of the island. The article ended with this: "This is the only deer on Nantucket Island, so Warden Jones has suggested to the state department that a mate be sent down to keep it company."[1] In 1926 Old Buck, as the swimming deer was called, finally had some company: two does were imported from Michigan.[2]

Nine decades later, in 2013, the *Inquirer and Mirror*'s May 19 paper ran an article entitled "Nantucket deer herd larger than estimated," which

---

* "gamesome wight," a playful person.

reported results of an aerial survey of the island's fifty square miles of land. The estimated number of deer was two thousand, or forty per square mile. The article continued with an account of the long-standing controversy about thinning the deer herd. Selectman Bob DeCosta was quoted as saying, "Here's the problem of why we're always going to have 2,000 deer on Nantucket: density. It's all posted land. Unless private property owners start letting hunters into their property to thin these deer out of these backyards, we can kill all we want in the middle of the island."[3]

In 2008 Nantucket had 411 cases of Lyme disease, up from 257 in 2007, among about 10,000 full-time residents and 40,000 summer residents. According to the CDC Nantucket was among the top three counties in the entire United States for Lyme disease on a per capita basis from 1992 through 2008. Other tick-borne diseases on the island are babesiosis and granulocytic anaplasmosis. In a survey of 220 households in 2009, 61 percent reported that one or more household members, including guests and renters, had had a tick-borne disease.

If Melville's tale of Moby Dick and the ill-fated crew of the *Pequod* is any measure, the whalers out of Nantucket of the 1850s had many dangers to face, but getting Lyme disease onshore was not one of them. Nantucket's residents and visitors now have more to fear on land than at sea. As do mainlanders. A few years ago, talking with a fellow plane passenger, I learned that she grew up around Lyme, Connecticut, in the 1930s and 1940s. She said that deer were so rare in the area then that a sighting of one merited an item in the local newspaper. That's hardly the case now, as deer rival squirrels in their numbers. The last time there might have been this many deer on the continent was before the last ice age (last glacial maximum), beginning about twenty-five thousand years ago. Arguably, there are even more deer now than during the preceding Pleistocene epoch, as most of the large predators, like big cats and wolves, have either gone extinct or been reduced to smaller, more confined populations since then.

The type of deer of interest here is the white-tailed deer (*Odocoileus virginianus*), which is the most widely distributed ungulate, or hoofed quadruped, in North America. This species' range includes much of the United States, southern Canada, Mexico, and Central America. The exceptions in the United States are Alaska, most of the Far West, including all of California, the Rocky Mountain region, and the Great Basin between the Rocky Mountains and the Sierra Nevada range. In these latter areas, the mule deer (*Odocoileus hemionus*) predominates. Its numbers have fluctuated too, but not nearly to the extent as those of white-tailed deer.

Although fewer natural predators might account for a higher overall density of deer than in precolonial times, there must be other explanations for the steepness of the rise in numbers. The deer population base from which the climb started was much lower than what it was before Europeans arrived. Now there are 30 million or more white-tailed deer in the United States—as many as one hundred per square mile in some places. In 1900 there were only a half a million total for the whole country. There had been a drastic reduction of the number of deer as the consequence of hunting and deforestation during the seventeenth, eighteenth, and nineteenth centuries in the eastern and midwestern United States. Forests were cleared for homesteads and agriculture and later for industry. Around 1900 there were fewer than two hundred deer in all of New Jersey. In Indiana, white-tailed deer disappeared entirely and were only reintroduced to the state in the 1930s. In Europe deforestation happened over a longer time and with generally greater respect for the husbandry of wildlife and preservation of forested spaces between towns and farms. But there, too, it had progressed to the point that outside of mountainous regions, deer were limited to small areas of woodlands.

In the northeastern United States, deer survived in a few places, sometimes in private hunting reserves on near-shore islands. Some deer, like the prospective mates for Old Buck, were brought in from undeveloped lands in the upper Midwest. After declining through the nineteenth century and the early part of the twentieth, forested lands made a comeback beginning in the mid-twentieth century. The center of gravity for agricultural activity in the United States moved westward. Eastern farms were abandoned or sold for residences, whose owners preferred trees and woodlands to expanses of grass or orchards. The landscape of the northeastern United States outside cities changed from one of open farmlands alternating with mills and other industries to deciduous forests. And as the forests returned, so did the deer. Because of "green belts" that extend from outlying forests into cities, deer are now found near the centers of such heavily populated areas as Philadelphia, Pennsylvania. Other factors resulting in the resurgence of deer were stricter hunting regulations: the decline of the human predators as well as deer's natural ones.

As Nantucket illustrates, landscapes that did not even exist in precolonial times can emerge. The bare sandy island described by Herman Melville in the mid-nineteenth century has since become covered with trees and shrubs. Whaling-themed tourist attractions have replaced whaling itself in the economy of the island. Where there once were no deer at all, there are now forty per square mile, far higher than the estimated ten to fifteen deer per square mile before the year 1600.[4] Whether some are descendants of Old Buck on Nantucket still, we can't know, but it's a good bet.

Lyme Disease in North America

Unknown or unrecognized on this continent before the 1970s, Lyme disease is now the most common arthropod-transmitted infection in the United States, and in the top ten overall among infectious diseases for which records are kept.* Its reported incidence—the annual number of new cases for a given population size—compares with new cases of HIV and *Salmonella* infections and outnumbers tuberculosis. Lyme disease is far more common than bubonic plague, typhus, or Rocky Mountain spotted fever, other bacterial diseases transmitted by insects or ticks. The rapid rise in incidence of Lyme disease in the United States from 1970 roughly tracks the dramatic increase in the populations of deer over the last several decades.

---

* Colds, the common upper respiratory viral infections, simple boils and related skin infections, and bladder infections far surpass other infectious diseases in frequency, but they are not individually reported as case reports to public health departments.

Lyme disease and erythema migrans were noted in North America several decades after they were first described in Europe.* This time gap cannot be attributed to the rash being overlooked by U.S. physicians, since the skin rash of early Lyme disease is distinctive enough that they would not have missed it. In fact, early medical reports about erythema migrans in U.S. citizens were only about cases who had acquired the disease while they visited or lived in Europe.[5] It was not until the late 1970s that erythema migrans began to be recognized in the United States.[6]

This does not mean that *Borrelia burgdorferi* was inadvertently imported from Europe on ships or airplanes to this continent, though that was an early theory after the pathogen's discovery. As discussed below, *B. burgdorferi* was undoubtedly on the North American continent before Europeans first stepped on its shores.[7] From descriptions of colonial forests and of ticks by explorers and the first settlers, it is apparent that conditions for *B. burgdorferi* transmission were present in parts of North America hundreds of years ago. Deer ticks in the northeastern United States declined in parallel with deer numbers. But small enclaves of *I. scapularis* hung on, such as on near-shore islands, on one of which it was observed in 1928.[8] *B. burgdorferi* was present too. Examinations of museum specimens of *I. scapularis* collected in Long Island, New York, in the 1940s revealed the presence of *B. burgdorferi* in some of the ticks, three decades before Lyme disease occurred often enough to attract attention.[9]

The inconspicuousness of the symptoms of early Lyme disease in comparison to those of infections such as smallpox, cholera, tuberculosis, and typhus, which affected many North Americans up through the nineteenth century, may have allowed Lyme disease to escape public notice when the American colonies were first developing and the ecological conditions for *B. burgdorferi* were favorable. The more consequential arthritic and neurologic manifestations of Lyme disease may have been occurring as well, but no connection was made between either of them and a skin rash or a tick bite. But the comparative mildness of early *B. burgdorferi* infection does not explain the continuing obscurity of the disease in

---

* Erythema migrans was sometimes called "erythema chronicum migrans" in medical literature in Europe, and this terminology was retained for a few years in the United States.

North America for the first three-quarters of the twentieth century. This was the time when medicine in North America and Europe was becoming rooted in the scientific method and was in its golden age of disease discovery and description. More deadly infectious diseases like smallpox were already in decline by 1900, and as stated, physicians in North America probably did not overlook the hallmark skin rash, erythema migrans. The inescapable conclusion is that Lyme disease is much more frequent now than it was in the first half of the twentieth century.

Life Cycle of *B. burgdorferi*

How did we get to this state of affairs, where Lyme disease has become a common infection in the United States and is listed by the U.S. National Institutes of Health as an emerging disease that merits high-priority attention? Some pieces of the jigsaw puzzle are fitting together: the correlation with the rise in the deer population, for one. But, as we saw in chapter 3, deer are not reservoirs for *B. burgdorferi*. So there are more pieces to put into place.

Before taking up again the emergence of Lyme disease and its current status, let's return to the story of its biology, left off at chapter 3's end. The three elements of pathogen, vector, and reservoir are linked together in a life cycle with the addition of a fourth critical element, the deer. A sustainable life cycle is just that: circular and repetitive. The cycle keeps turning, maintaining if not spreading the pathogen in nature. What figure 4.1 depicts is drawn from the ecology in the northeastern and north-central United States and features the pathogen *B. burgdorferi*, the reservoir white-footed mouse, the vector *I. scapularis*, and the adult tick's primary host, white-tailed deer. There are some differences from life cycles of Lyme disease agents in the Far West of North America and in Europe, but the life cycles are close enough for this example to be relevant to other *Borrelia*-vector-reservoir cycles as well.

In brief, here is the cycle in general terms, arbitrarily starting with a larval stage tick acquiring the microorganism from an infected vertebrate. That infected tick in its next stage as a nymph bites a vertebrate, thereby

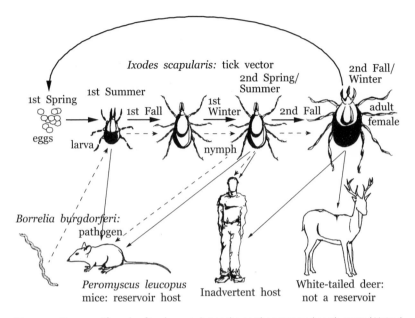

Figure 4.1. Two-year life cycle of *Ixodes scapularis* in the northeastern and north-central United States. There are four life, or developmental, stages for hard-bodied ticks such as *I. scapularis*: egg, larva, nymph, and adult. Transitions between stages are indicated by thick solid lines with large arrowheads in the picture; the usual time of year during which each change occurs is given above the arrow. The cycle for an individual tick begins in the spring of the first year, when an adult female tick who had fed on a deer during the fall or winter lays her eggs. The larva emerges from the eggs in the spring or summer and feeds on a mouse, possibly becoming infected with *Borrelia burgdorferi* from the mouse. (Feeding is indicated by thin solid lines with small arrowheads.) After feeding, the larva changes to the next stage, a nymph, during the fall. The nymph does not usually feed until it has passed through the first winter. In the spring or summer of the second year the nymph feeds on a small mammal or bird or sometimes a human (an inadvertent host). After its blood meal, the nymph changes to an adult in the late summer or fall. The adult female feeds during the fall, winter, and sometimes the early spring on deer and occasionally on other large mammals and humans. During feeding, the adult female mates with a male, who does not feed during this stage. The female produces eggs and then dies. The two-year cycle begins again with the eggs.

Superimposed on this cycle for the ticks is the life cycle of the pathogen, *Borrelia burgdorferi*. There is no transmission of the pathogen from an infected female to its offspring through the eggs. The ticks first become infected as larvae when they feed on an infected *P. leucopus* mouse or other animal that serves as the natural reservoir host for the pathogen. (The direction of infection in the cycle is indicated by thin dashed lines with small arrowheads.) Once infected, the tick continues to carry the pathogen as it goes through the different stages. An infected nymph will transmit the infection to other mice and small animals in the spring or early summer, thus providing a new set of mice to infect the larvae. Although adult ticks are also capable of transmitting infection to small animals or to humans, the contribution of adult tick bites to human disease is much less than for those of nymphs. Deer are largely resistant to the infection and do not serve as reservoirs. Drawing by Nathan Barbour and Alan Barbour.

passing the infection on. An uninfected larva then bites this infected vertebrate and so on. For some tick-borne infections, the pathogen can be transmitted vertically, that is, from an adult female tick to its offspring through the eggs. This provides an alternative for the pathogen to persist in an environment across generations. But neither *B. burgdorferi* nor any of the other Lyme disease *Borrelia* species allow this option. Their life cycles depend on transmitting pathogens with each generation of ticks to a new susceptible host among vertebrates in the area.

I. *scapularis* ticks that transmit Lyme disease live about two years. A larva, which is not much larger than a poppy seed, hatches in the summer from an egg laid by an adult female that spring after making it through the winter. The larva climbs up a blade of grass, onto leaves of a low shrub or on top of the leaf litter, and waits, poised to attach to a passing animal. It has crawled away from its siblings that have hatched, but for a creature this small, it would not be far. If lucky, the larva attaches to and feeds on a small mammal or bird that summer or early fall. If that host is infected, *B. burgdorferi* may be taken up by the feeding tick. The blood meal provides the energy and nutrients for the larva to molt to a nymph during that first year and then for that nymph to pass through the first winter without feeding. A proportion of the spirochetes that entered the larva and then multiplied will pass through the molting process to the next stage as well.

Unfed nymphs are about the size of a sesame seed. The nymph may survive the winter with its subfreezing temperatures and aridity by entering a state similar to hibernation in which metabolic processes slow down. Ground cover or leaf litter provide the insulation and moistness necessary for surviving winter, but many nymphs will die. In late spring of the second year, the nymph becomes active once more and "quests," or seeks out a host for its sole blood meal. This again is by lying in ambush on low-growing vegetation. When it is hot and dry in the afternoon, the nymph may retreat into the shade and higher moisture of leaf litter or tall grasses to avoid dehydration, and then reemerge the next morning when dew may be present. The preferred hosts, such as white-footed mice, voles, shrews, chipmunks, and ground-foraging birds, are the same hosts that larvae feed on. Occasionally, the tick's source of blood is a human or a dog. This is unfortunate all the way around. It's a bad outcome as far as the pathogen and tick are concerned, because these hosts are dead ends for further transmission or a ride back to the forest. And the damage might have been done for the person or their pet. Most of the cases of Lyme disease in the northeastern and north-central United States are from the bite of a nymphal tick.

When a nymph that was previously infected as a larva is feeding on a new host, the number of spirochetes in its intestine may grow into the thousands. The nymph may also become infected for a second time from the host it's attached to. But in the end, only a small fraction of the *B. burgdorferi* cells successfully make the journey to the salivary glands. That's why the odds of a human getting Lyme disease from the bite of a single infected tick, even if embedded for two or three days, are no more than one in twenty. The reason the white-footed mouse and some other important reservoirs are so much more frequently infected is that many more than ten ticks cumulatively attach and feed over the weeks of high questing activity for the nymphs.

After the blood meal, the nymph molts to an adult and is revealed as either a female or a male. If the tick had been infected as a larva or nymph, spirochetes almost always pass through to the adult stage. The adults climb higher up on the vegetation to better grab hold of deer and other large or medium-sized mammals as they stand or walk by. Although deer are the host of choice for adult *I. scapularis* ticks, they are not very susceptible to *B. burgdorferi* and are not suitable as reservoirs. The adults feed later in the year than nymphs and larvae and may remain active even as temperatures drop to just above freezing in the late fall.* If a female adult tick does not attach to a deer or other suitable mammal and complete its blood meal while simultaneously attracting a male infesting the same animal, it will not be able to produce eggs the next spring. If there are no deer or if deer are in much reduced numbers, in the absence of a suitable alternative for the adult ticks, the numbers of *I. scapularis* ticks in the area will decline.

---

* People occasionally present with early Lyme disease in the fall or early winter in the northeastern or north-central United States; this is usually from the bite of an adult tick.

For *B. burgdorferi*'s benefit, the larvae should feed on the same types of hosts as nymphal ticks and after the nymphs have fed. If the feeding activity periods for nymphs and larvae are simultaneous or substantially overlap, larvae may still get infected, but there would be fewer infected rodents to choose from. If either the nymphs feed after the larvae or larvae and nymphs feed on different types of hosts, then there is little chance for *Borrelia* to be maintained in an environment.

In the far western region of North America, larvae tend to feed during the same period as nymphs if not before them, and this is one explanation for the lower prevalence of infection of *I. pacificus* ticks there. In both far western and southeastern regions, immature *I. pacificus* or *I. scapularis* ticks feed on the abundant supply of lizards available. But lizard blood is lethal for any *B. burgdorferi* cells that may be present in the tick's intestine, thus further reducing the odds of being bitten by an infected tick. Even if infected nymphs are to be found in a far western or southeastern environment, this stage of tick, which is so important for transmission elsewhere in North America, is seldom encountered by humans.

On the Eurasian continent, even though two other Lyme disease *Borrelia* species dominate, the life cycles for the pathogens are basically like that of *B. burgdorferi*, and the same principles hold. But the species of reservoirs among mammals and birds for the pathogens as well as those of the hosts for the immature and adult ticks are largely different from those we see in North America and can also differ between two Eurasian *Borrelia* species. For instance, *B. garinii* tends to use birds as reservoirs, while *B. afzelii* uses small mammals.

Climate and Landscape

The enduring success of a life cycle for a tick-borne pathogen is contingent on environmental and climatic conditions. For sure, the environment needs to support the tick itself, which has its tolerances for temperature and humidity. One of the reasons for the high density of *I. scapularis* in coastal regions of the northeastern United States since the end of the twentieth century has been the moderate temperatures and humidities in both winter and summer, when compared with areas farther inland or at higher altitude. Deer ticks can survive frigid conditions, but not for extended periods. The ticks, which feed and drink but once a year, become dehydrated under prolonged dry conditions, and when moisture levels become too high, they are susceptible to mold and other fungi. Temporary refuges from the weather and sunshine are leaf litter, long grasses, and low-lying vegetation.

For both *B. burgdorferi* and the tick, the environment and climate also need to be conducive for the white-footed mouse (or other rodent filling this role), as well as a variety of alternative small animal reservoirs and hosts. The numbers of each of these vertebrates in turn depend on the supplies of food, which range from seeds and nuts to insects and other invertebrates, as well as on the numbers of predators, parasites, and pathogenic microbes. Predators include foxes, coyotes, weasels, hawks, owls, and snakes. Parasites include intestinal worms and flies that lay eggs under the skin of the mice. Pathogenic microbes include viruses that can be deadly for the mice but hardly affect humans. To sustain generation after generation of ticks, large mammals need to be on hand as blood meal sources for the adult female. These typically are deer, but domestic and other ungulates, like cows, sheep, and wild boar, might serve as well, as the experience in Europe shows.

Ixodes ticks that transmit Lyme disease are forest- and moisture-loving ticks. But for conditions to be right for *B. burgdorferi*'s life cycle and then to put as many humans as possible at risk, there are other desirable features for a prospective habitat. These features were first noted in the coastal Northeast of the United States where *B. burgdorferi* reemerged. A suitable landscape is a "disturbed mixed forest," which means second-growth forests of deciduous and evergreen trees, with saplings, leaf litter, and sufficient sunlight to support growth of shrubs and other low-lying vegetation beneath trees. Other fostering habitats are old farm fields that are reverting naturally and transition zones between grasslands and brush. These are characteristics of suburban and rural areas undergoing reforestation in the northeastern and north-central states as well as bordering areas of Canada. Visualize a house with a backyard that abuts a woodlot with dense shrubbery at lawn's edge, or a grass-bordered footpath meandering through a long-abandoned orchard, shared by hikers and deer, and you get the idea.

The geographic ranges, or distributions, of the vector *I. scapularis*, the primary reservoir *P. leucopus*, and white-tailed deer are each considerably larger than the areas occupied by *B. burgdorferi* currently. The white-footed mouse is found farther north, west, and south of the high-risk areas for Lyme disease. The white-tailed deer occurs throughout much of North America, including areas where *B. burgdorferi* has yet to be found. The deer tick's range extends down into the southeastern United States, but *B. burgdorferi* is uncommon in the ticks there. Figure 4.2 shows this graphically. In a four-year study in which ticks were systematically collected each summer, *I. scapularis* nymphs were found across the eastern United States, from the north into the south.[10] But within the larger region, areas where the ticks had *B. burgdorferi* in them were more restricted.

Although we can justifiably point to the rise in deer populations as a cause of the emergence of Lyme disease in North America, in truth we know that other conditions have to be in place as well for *B. burgdorferi* to prosper. This includes the right timing of larval and nymphal tick feeding and the availability of suitable hosts. Even though these conditions in combination explain why Lyme disease risk is largely limited to certain well-defined areas, we should not be complacent that these are forever its geographic limits. History of the pathogen teaches us not to take this for granted.

The Reemergence and Spread of *B. burgdorferi*

Best evidence places *B. burgdorferi* in North America from prehistoric times, at least since the last ice age. This inference holds for the east and west of the continent and its temperate latitudes. The general exceptions have been high mountains, prairie grasslands, deserts, and subtropics. In the Far West, there has been less fluctuation of *B. burgdorferi*'s populations over time than elsewhere, because it was less subject to deforestation, habitat loss, and deer hunting in the nineteenth and early twentieth

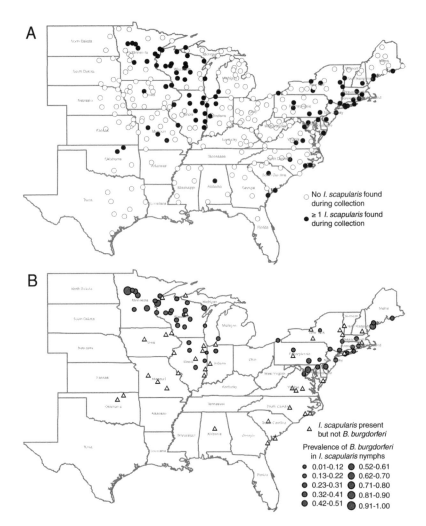

Figure 4.2. A, The locations of state parks and other parks where ticks were collected in the eastern United States during the summers of 2004, 2005, and 2006 are shown as circles. Open circles indicate that no questing *Ixodes scapularis* ticks were found at the sites. (A questing tick is one that is actively seeking a host to feed on.) Closed circles indicate that at least one questing *I. scapularis* tick was found at the site. **B,** At the locations where *I. scapularis* were found, the presence of *Borrelia burgdorferi* in each of the collected ticks was determined. The locations with ticks without infection are indicated by open triangles. Locations with infected ticks are indicated by gray-filled circles, the diameters of which are proportional to the frequency of infection (or prevalence) at each site. The distribution of infected *I. scapularis* ticks corresponded with the distribution of Lyme disease cases (see next figure).

The full methods and results are given in M. A. Diuk-Wasser et al., "Human Risk of Infection with *Borrelia burgdorferi*, the Lyme Disease Agent, in Eastern United States," *American Journal of Tropical Medicine and Hygiene* 86 (2012): 320–327. Map and graphics by Alan Barbour and Tony Soeller (University of California Irvine).

centuries. Present-day strains of *B. burgdorferi* in Northern California are substantially different in makeup from those found to the east, an indication of a different back story for the pathogen in the Far West.[11]

In the eastern half of the United States, two overlapping populations of *B. burgdorferi* are traceable to a time thousands of years before European settlements. The center for one population was in the Northeast. The second, which appears to derive from the first and not vice versa, was centered in the north-central region. As previously told, these populations were much reduced in size and scope as those parts of the country developed agriculture and industry, and as deer were hunted to near elimination. Relicts of both populations persisted, such as on islands in the case of the northeastern population. These refuges served as sources of *B. burgdorferi* when conditions changed with reforestation and deer populations rebounding and beyond.

The landscape and other environmental changes have been most dramatic in the northeastern states. The tick population of Westchester County, just outside New York City, steadily increased over several years beginning in the 1970s. There was an independent expansion of the populations of *I. scapularis* ticks and *B. burgdorferi* in the north-central United States. *I. scapularis* was first noted in Wisconsin in 1970, three decades after its observation in the Northeast.

In both of these large regions, *I. scapularis* and *B. burgdorferi* have continued to extend their range after their reemergence. This can be seen in figure 4.3, which shows the five-year incidences of Lyme disease by U.S. county during two periods: 1992 to 1996 and 2007 to 2011. From the coastal areas, the ticks moved inland: up the Hudson River valley and the Connecticut River valley, from Cape Cod westward and farther into Massachusetts, and from coastal areas of New Jersey to the west into Pennsylvania. The ticks are found at northern latitudes at which they had not previously been seen in human memory in the United States. Lyme disease now is reported in the northernmost parts of New Hampshire, Vermont, and Maine, as well as in the southernmost areas of the adjacent provinces of Canada.[12]

The north-central expansion from the border area of Minnesota and Wisconsin has been into the remainder of those states and into Michigan,

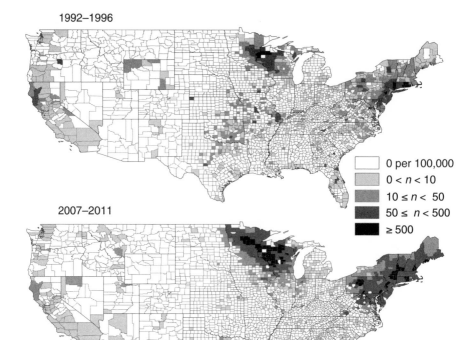

1992–1996

2007–2011

0 per 100,000
0 < n < 10
10 ≤ n < 50
50 ≤ n < 500
≥ 500

Figure 4.3. Lyme disease is spreading in the United States. The figure shows the five-year incidences of Lyme disease by U.S. county during two periods, 1992–1996 and 2007–2011. The rates are per 100,000 population in the year 1990 for the 1992–1996 period and in the year 2005 for the 2007–2011 period. Counties with fewer than two reported cases during the five-year period are white in each map. Counties with at least two reported cases in a given period are shown in shades of light gray to black, on a scale corresponding to the incidence in the county. The presence of a case in a county does not necessarily mean that there is transmission of Lyme disease in that county. The infection may have been acquired by a resident at another location; case reports may not list the county of acquisition. The overall trend between the two periods was of a spread of Lyme disease to new areas from its original locations in the northeastern and north-central United States. The apparent decrease between the two periods in the number of cases over a geographic region that stretches diagonally from north Texas through Oklahoma and into Missouri is probably due to the greater recognition during the interim that these were cases of southern tick-associated rash illness (STARI) rather than Lyme disease (chapter 11).

The publicly available data on reported cases of Lyme disease by United States counties are from the Centers for Disease Control and Prevention, and the population by county data are from the United States Census Bureau. Map and graphics by Alan Barbour and Joanne Christopherson (University of California Irvine).

Illinois, Indiana, and North Dakota. While Lyme disease's reach has extended southward as well, namely into Maryland, Delaware, and Virginia, it has not been as far reaching as the northward and westward movements. In the Far West, there has not been much change in either Lyme disease incidence or its extent. In the south-central part of the country, for a band stretching from north Texas through Oklahoma into Arkansas, there were fewer reports of Lyme disease in 2007 to 2011 compared with the period fifteen years previous. (This apparent change is discussed in chapter 11.)

These range extensions are in part attributable to reforestation and the increase in deer populations. Another factor that may have facilitated the movement of ticks into areas now conducive for them is the migration or other travels of birds that are infected themselves or carrying infected ticks. But global warming also accounts for the northward expansion.[13] Adult and nymphal ticks are more likely to survive the winter when temperatures moderate. Sweden has likewise observed a northward extension of the range of *I. ricinus* ticks toward the Arctic Circle,[14] but this extension cannot be attributed to reclamation of agricultural lands for forests.

Risk Factors for Lyme Disease

What is the single best predictor for getting Lyme disease? It is having had Lyme disease in the past.[15] That answer sounds like the punchline of a joke, but it's no laughing matter for people at continued risk. This one fact shows that *B. burgdorferi* infection is not the proverbial bolt from the blue, randomly affecting a person who happens to be in the general area where Lyme disease occurs. Within, say, Westchester County, New York, a largely suburban area with a comparatively high incidence of *B. burgdorferi* infection, there are some living situations and activities for which the odds of getting Lyme disease are low to none, and others for which the risk is substantial and recurring.

But before exploring those factors, we should remind ourselves that there is really only one way to get Lyme disease in North America: let a deer tick that is carrying *B. burgdorferi* feed on you for at least a full day. A small risk attaches to working with *B. burgdorferi* in culture in a laboratory or to handling infected small mammals, but few people are doing that. There are no proven instances of transmission from one person to another through sharing a toothbrush or household items, intimate or casual physical contact, blood transfusion, organ or tissue transplant, or by breast milk. Hunters in the fall may be at some risk, but not from handling or dressing a deer's carcass. Rather, they risk inadvertently picking up unembedded adult ticks from the deer.

So it's down to ticks. Ask people on the street where someone might pick up a tick, and the answer might be "in the forest," "in a wilderness region," "on a hike," or "on a hunting trip." Ticks are often thought of as hazards only for people away from home. They are not generally considered to be domestic threats. Although many people do acquire their *B. burgdorferi* infections out of town—perhaps hiking, camping, fishing, or hunting—most cases of Lyme disease in North America begin with an encounter with a tick living around or close to human residences. This is called a "peri-domestic," literally "around the home," situation and could be a yard for a single family home, a wooded or shrubby area abutting a multiple-housing-unit dwelling, or a neighborhood park. Characteristically, the people who live in the area know that deer are frequently found in the location or in an adjacent area, and bites by deer ticks are a recognized nuisance if not a hazard. The greater the number of hours spent in these environments, the better the odds of infection.

In a Massachusetts town, the odds of becoming infected at or near home were directly related to the distance between the home and a forested nature preserve. A field study in suburban Westchester County showed that while most ticks were located in wood lots and in unmaintained edges between forest and lawn, there were also ticks on ornamental plants and in the lawns of people's homes. Two-thirds of all the lawns with Lyme disease tick vectors in the study had adjoining wood lots. A study in New Jersey showed that children with Lyme disease, in comparison with their disease-free neighbors, were more likely to play in their backyard than inside the house or in the front yard. Front yards generally have well-trimmed grass; backyards more often abut a stand of trees or dense shrubs.

Some people might not live in a residential area with these features, but their jobs or recreations may put them in harm's way in parks and other natural areas where Lyme disease *Borrelia* occur. Forestry workers in the Netherlands and England have a higher risk of getting *B. burgdorferi* infection than others living in the same areas. Soldiers in Austrian woods and hikers in Swiss mountain areas have about a one in ten chance of becoming infected over the course of a summer outdoors. In Sweden, people with weekend and summer homes in an area outside Stockholm were two to three times more likely to have had *B. burgdorferi* infections than were those who tended to stay in the city. Investigations of Lyme disease cases in California revealed that the risk of disease went up with the number of hours spent hiking on trails and pursuing other outdoor recreations. The outdoor areas with lower risk are farm lands, athletic fields, and beaches, but many of these areas are adjacent to wooded areas. The fairways of the country club may be safe, but there is greater danger than a high score from a ball hooked into the deep rough or the adjoining woods.

A count of Lyme disease cases by gender and age in the United States shows that cases are about equally divided between males and females. The age groups with the highest incidences of infection are ages 5 to 9 and 50 to 54. Why those particular ages? My hunches are that children in that age range are more likely to play outside than either younger or older cohorts, and that adults of those ages may have achieved occupational and personal goals by that age and have more time for hobbies and other recreational pursuits around the home and nearby.

The time of year also figures in an estimate of the odds of infection. Most cases of Lyme disease are acquired between mid-May and mid-July in the northeastern and north-central United States. This period corresponds to highest activity for the nymphal *I. scapularis* ticks in those regions. Unfed larval ticks pose no threat, but there is some risk from adult ticks in the fall and early winter.

Finally, we need to distinguish between being at risk of an *Ixodes* tick bite per se and of also getting Lyme disease or one of the other infections transmitted by these ticks (chapter 11). The tick bite itself is to be avoided, not only because it's an unpleasant nuisance but also because it could lead to a staph or strep skin infection. And as we saw, some people develop a type of allergy or hypersensitivity to the tick's saliva that can produce swelling and inflammation around the bite. In many areas of North America and Eurasia, that's about all that one need fear. But in others, like the northeastern United States, spirochete infection rates in nymphs average 25 percent and can be as high as 40 percent at some sites. Intermediate in risk are areas like Northern California, where the infection rates in *I. pacificus* ticks seldom exceed 5 percent.

Epidemiology of Lyme Disease

With a fuller appreciation of the ecology and history of *B. burgdorferi* and the risk factors for infection, let's pick up again the story of Lyme disease in North America with a more in-depth look on where it currently stands. The first stop is an accounting of the number of cases of Lyme disease. This is the domain of epidemiology, which is the science of disease and health in populations. Epidemiologic studies are valuable for tracking the spread of disease and for identifying who is at risk of disease. These studies depend on accurate counts of disease throughout much if not all of a country. Local, state or provincial, and federal public health agencies need not tabulate every case or even most cases to gather informative data, as long as the counting is a representative sample and commensurate from place to place and year to year.

One practical benefit of epidemiology is that it assists in the diagnostic process by providing a reasonably accurate answer for use by the doctor who asks, "What is the likelihood that the patient before me has Lyme disease?" As with other zoonotic (transmitted by animals) and vector-transmitted infections, a story of exposure to an environment in which infection transmission occurs is a very important clue for assessing the odds of a person's having the disease. If there were wider appreciation by a community's physicians and its inhabitants that *B. burgdorferi* transmission in their area was either impossible or extremely unlikely, the number of misdiagnoses would decline. Consider that the diagnosis of malaria is seldom made in South Dakota, because physicians and public health workers recognize that there is no transmission of that mosquito-borne parasite in that state or in the adjoining ones.

If these epidemiologic data are to be reliable and reproducible, however, strict case definitions, upon which most investigators agree, are needed. The reasoning here is that it is better to err in the direction of missing some true cases of a disease than to err in the direction of including illnesses that are not in fact the disease under study. If too many false cases of Lyme disease are included in the studies, for example, then the research and epidemiology data may be distorted, and as a consequence, predictions about the impact and spread of the disease may be inaccurate. Another rationale for the use of strict case definitions is that it enhances the accuracy of clinical trials of new diagnostic tests and new therapies. If a laboratory test or treatment works for cases that are conclusive by strict criteria, then it is likely that it will work for a case of the same infection that falls short of the strict case definition.

The main arbiter of the case definition for epidemiologic records in the United States is the federal Centers for Disease Control and Prevention, commonly known as the CDC. In actual practice the CDC accepts the consensus case definition of the Council of State and Territorial Epidemiologists. The CDC has gone through several versions of its case definition for Lyme disease over the last three decades, but they have more similarities than differences. As of this writing, the latest version dates from 2011. Some of the essential features can be summarized thusly:

1. If there is an erythema migrans–like rash, it must be at least 2 inches (5 centimeters) across its largest diameter and observed by a physician. Self-reporting of a rash by a patient is not enough.
2. An erythema migrans rash so defined must have had an onset within thirty days of exposure to "woody, brushy, or grassy areas" in a county in which Lyme disease is known to occur. A tick bite need not be noted. But for a county to meet this "known" criterion, it should have either two confirmed cases of Lyme disease that have been acquired in the county or demonstrated *B. burgdorferi* in *I. scapularis* or *I. pacificus* ticks collected in the county.

3. If there is an erythema migrans rash but not a known exposure, there must be laboratory confirmation, either by a positive culture for *B. burgdorferi* or by demonstration of antibodies to the pathogen.

4. If there is no documented erythema migrans, there should be at least one late manifestation of joint, nervous system, or heart symptoms of Lyme disease, plus laboratory confirmation. The laboratory aspects for diagnosis are discussed in chapter 6.

Case reports of Lyme disease are usually acquired passively by the state or local public health agency. That is, reporting is dependent on physicians filling out the form and sending it in by mail, by fax, or online. Sometimes agencies will do a more active form of case counting through outreaches to physicians and clinics, but these activities may not be sustained. And the phenomenon of "reporting fatigue" may set in, especially after a period of more intense public interest and concern about Lyme disease has passed. While Lyme disease is a federally designated "notifiable" disease, it is no secret that the public health implications of a failure to report are not as great as for a new case of tuberculosis or measles in the community, and there is little or no enforcement of reporting requirements at the medical practitioner level.

All this undoubtedly adds up to an underestimation of the true number of cases of *B. burgdorferi* infection. This situation is not unique to Lyme disease; there is no conspiracy to underreport Lyme disease only. This circumstance applies to several other zoonotic infections that do not pose a direct public health threat through person-to-person transmission, unlike tuberculosis, measles, or syphilis (which do pose such a threat), or threats that may have terrorism ramifications, like anthrax.

Before we consider the magnitude of the undercount and look at attempts to adjust for it, let's accept the official accounting of cases and see where it stands now. The annual number of confirmed reported cases of Lyme disease in the United States is about 30,000. This is a dramatic increase from 1982, when reports of cases began to be tabulated, and since 1991, when reporting was mandatory. Overall for the entire country, there are currently about 8 reported cases per 100,000 people each year. But the incidence rates vary greatly by location and are concentrated in particular regions. In 2012, 95 percent of the confirmed cases of Lyme disease in the United States were from thirteen states: Connecticut, Delaware, Massachusetts, Maryland, Maine, Minnesota, New Hampshire, New Jersey, New York, Pennsylvania, Vermont, Virginia, and Wisconsin. In many counties in these states, the incidence exceeded 100 per 100,000 people. On Nantucket in 2008, it was close to 1,000 per 100,000, or 1 in every 100 people that year.

Using additional sources of data for Lyme disease diagnoses besides public health case reports, the CDC estimated that the actual count of new cases of Lyme disease in the United States is closer to 300,000 annually,[16] roughly ten times the number of official cases. This adjustment was based on examination of medical claims for individuals enrolled in health insurance plans and the results of laboratory tests for Lyme disease that were done at several large commercial laboratories.[17] On Block Island, Rhode Island, in 2009 and 2010, there were ten cases of Lyme disease for every one that the state health department accepted reports for.[18]

Block Island is 13 miles (21 kilometers) south of the Rhode Island coast and about halfway on a sail from Nantucket to Shelter Island at Long Island's tip. The island is a remnant of the retreat of the last glaciers that covered much of the continent. It is part of a chain with Nantucket, Martha's Vineyard, and Long Island. In the nineteenth century, it was described as a "bleak storm-swept island."[19] Excavations of sites dating to around 1000 BCE revealed that the animal component of the diet of the precolonial native peoples consisted primarily of shellfish, migratory water fowl, fish, and seals.[20] At these prehistoric sites were some remains of white-tailed deer, but they appear to have been only a small part of the native diet. At some point deer were eliminated entirely and did not return until the 1960s, when a few were shipped to the island. By 2014 the herd had grown to between eight hundred and one thousand deer, or one for every permanent resident of Block Island.[21]

Chapter 5
Approach to Diagnosis

Scene 1. Put yourself in a doctor's shoes. It is mid-July on Cape Cod, Massachusetts, and a 45-year-old patient sitting across from you is telling her story. The illness started ten days ago with what the patient attributed to a "summer flu": chills, muscle aches, a headache, and tiredness. But she is not getting better. You learn that the patient and family live in Iowa but were renting a vacation house near the shore for June and July. No one else in the family of four had been recently ill. The patient had previously been healthy and was not taking any medications. While the family had seen and removed small ticks from their clothes, no one had noticed a tick embedded in the skin. On physical examination, the patient has a low-grade fever, and high on the back of her left thigh is a 5-inch diameter rash, reddest at its well-demarcated outer edge and fading toward the center. There is an enlarged lymph node in the left groin. Are you ready to give the patient's illness a diagnosis or do you want some more information, such as a blood test result?

Scene 2. You are still the doctor, but a different patient is sitting across from you, and now it's February in Los Angeles. On the face of it, this patient's illness sounds similar to the one of Scene 1: recurrent pains of joints and muscles, headache, easy fatigability, and difficulty concentrating. But it started two years ago, not ten days. The patient associates the illness with a hiking trip in the Rocky Mountains two months before the illness's onset. A tick bite had occurred during the hike. On physical examination, his body temperature is normal, and there is no rash, joint swelling, nerve weakness, or other abnormalities. Other physicians have been consulted over the last several months for the illness, but "all the tests and x-rays were negative." The patient has been reading a lot about Lyme disease on websites and on Facebook and is convinced that he has that illness. Are you ready to assent to the patient's diagnosis for the illness and provide the prescription for antibiotics the patient is asking for? Or do you want some more information, such as a blood test or an x-ray result?

The example of Scene 1—an ill person goes to a physician and places herself in the doctor's hands for diagnosis, education, and treatment— is still a common medical encounter. But increasingly, with the huge amount of information and opinion that can be accessed and shared through the Internet and social media, people feel more empowered to make a diagnosis themselves, as in Scene 2. Whether the patient passively receives the diagnosis or actively proposes it (or some combination of the two), the answers to the questions about diagnosis posed above are not easily answered with a simple "yes" or "no," as we will see through an exploration of the basics of the diagnostic process.

Putting a Name on It

As Aristotle recognized, humans are prone to sort and categorize. Various peoples and cultures have classified animals and plants by groupings of common traits. Some of these groups, such as classifying whales with fishes, turned out to be erroneous. But there was usually some logic in these early classification attempts, such as the water-dwelling characteristic of whales and fishes. These different sortings of living things into categories eventually led several centuries later to the universally adopted system of naming we have now: each type of organism is assigned a genus name followed by species name, for example, *Homo sapiens* for modern humans. At each higher step in the taxonomy, the groups become larger in size and more inclusive: first primates, then up to mammals, then up to vertebrates, and so forth. When confronted with an animal or plant never been seen before, there is a strong inclination to place it where it best fits among existing types of animals or plants. As we saw with

Borrelia bacteria in chapter 3, a newly discovered organism may be judged similar enough to a known species to add it to the list of examples of the species, or there may be sufficient justification for a new species designation if not a new genus. But just to leave it as a "mystery creature" or something equally vague would never do.

Clinicians—physicians who see patients—are just as constitutionally inclined to classify as zoologists and botanists are. Physicians want to put a name to whatever is causing the symptoms of the patient facing them. It may be enough for most of us to know our head hurts when we go to the pharmacy for an over-the-counter medication. We need not know the biochemical explanation for the headache. For much of the history of medicine, symptomatic relief was the most one could hope for. But even when the understanding of disease processes was limited and few effective therapeutic options existed, choosing among those options was not as random the cast of a die. It depended, for better or worse, on the medical practitioner following some scheme for organizing illnesses into different categories.

Now there are a dizzying number of treatment options, including the old standbys of a pain pill or reassurance alone. Symptomatic relief may be the primary objective for both patient and medical practitioner, and in the end, that may the outcome that counts most. But unless the care's cost is entirely borne by the patient and physician—that is, without payment or reimbursement by a third party—a medical visit leading to a prescription or hospitalization has to be linked in writing or electronically to a medical diagnosis or diagnoses. This is the expectation for sure when the therapy is aimed at either curing the disease or directly ameliorating its pathologic effects. It is usually not acceptable to health insurance companies or government health agencies (such as Medicare in the United States) for a practitioner to prescribe a medication or perform a procedure without an explicit, specific, and defensible reason.

Not surprisingly, there are rules and standards for the names and criteria for medical diagnoses. A diagnosis of a bone cancer has to mean the same thing in Kansas as it does in Oregon if there is to be consistency of care and prognosis and if costs and outcomes are to be commensurate. The formal designations for diseases are increasingly standardized across countries and continents. At last count, there were more than 68,000 individual codes, with names like R53.82 and M79.7, for medical diagnoses in the current version, the tenth, of the International Classification of Diseases (ICD-10).[1] Several codes apply specifically to Lyme disease. These include "erythema migrans due to *Borrelia burgdorferi*," "meningitis due to Lyme disease," "arthritis due to Lyme disease," and "myopericarditis due to Lyme disease," as well as the catchall "Lyme disease, unspecified." There are specific codes for "chronic fatigue syndrome," "postviral fatigue syndrome," and "fibromyalgia." There are not, as of this writing, codes for "chronic Lyme disease" per se or for "post–Lyme disease syndrome" (chapter 10).

As these few examples show, some codes appear to require either evidence of the pathogenic microbe's presence (for example, *Borrelia burgdorferi*) or of a particular pathologic change (for example, meningitis). Other codes, such as for chronic fatigue syndrome, may not call for anything more than a particular constellation (a syndrome) of symptoms without the proof of an offending microbe or pathologic change that one would expect to see under the microscope. This distinction between the two general types of information upon which to build a diagnosis I consider next.

Symptoms and Signs

The Enlightenment philosopher John Locke distinguished between "primary" qualities and "secondary" qualities of a thing, such as a stone, rose, biscuit, or hammer. While a primary quality is independent of our perception of it, or our existence at all, a secondary quality is dependent.* Examples of primary qualities are weight, length, and density of an object. These may have different names for a Martian, but what we, as Terrans, measure and what Martians measure would be the same, and would be so whether Terrans or Martians ever existed. If there were an

* The primary-secondary dichotomy is akin to the objective-subjective distinction. But the two latter words have so many meanings, we profitably step back and frame this in other terms.

English-Martian dictionary, we would see that we agree on the mass of an object, like the stone, or the length of a hammer. Examples of secondary qualities are color, taste, and smell. What we experience as the "pink" of the rose would be a shade of gray to a dog, though the wavelengths of the light reaching human and canine eyes from the flower are the same. Taking some liberty, we can extend this concept to a disease. Some primary qualities would be the patient's temperature in some standard of units like degrees, the volume of fluid in the swollen joint, the concentration of an inflammatory cytokine in the blood, the strength of the kneecap reflex, the count of white cells in a dysfunctional organ, and the number of pathogen cells in the body. If I am the patient under exam, all of these exist independent of my consciousness. I could be in a coma, and my leg would still involuntarily extend when the kneecap is tapped, and the same number of white cells would circulate in my blood.

When we are both the perceiver and the stuff perceived, we are in the realm of secondary qualities. These include such everyday sensations as hunger and thirst as well as what we commonly refer to as "symptoms," such as chills, headache, pains in the joints, and breathlessness. The sensation of chilliness may correspond to the primary quality of temperature as measured by a thermometer, and the joint discomfort may correspond to primary qualities of the circumference of the knee and the angle to which the limb can be flexed, but they are not equivalent. The temperature and joint circumference we commonly call "signs." They can be observed directly by another person. We can empathize with someone in pain from arthritis, but short of being Mr. Spock from *Star Trek*, who can perform a "mind meld" with another person, we do not know for sure what the pain is really like for that other person, even if we have experienced something like that joint swelling ourselves. One may grade one's own headache severity on the 10-point pain scale ubiquitous in hospital emergency rooms, but my scale may have a different baseline for 1 than someone else's.

Since the ancients laid their hands on a feverish person's forehead and scrutinized sensible properties of the urine, physicians have generally preferred to base a diagnosis on primary qualities—that is, signs—either directly observable in the patient or as revealed by some apparatus or laboratory test. With aid of instrumentation, the body temperature can be determined precisely, the white cells in the blood can be counted under a microscope, cancerous cells can be distinguished from normal cells, the binding of antibodies to proteins can be measured with chemicals, and the pathogen can be isolated from the patient and grown in the laboratory. It is there to see in the culture tube. We can sequence all or parts of its DNA to tell us exactly what species it is. These results each would have a common reference for medical practitioners, whether in New York, Shanghai, Paris, or Nairobi. In a broad sense, and in keeping with the major distinction between primary and secondary qualities, these procedures and test results can be considered "signs" as well.

Medical science is finding primary quality correspondences for more and more secondary qualities. Technologies such as functional MRI scans can identify which parts of the brain are lighting up when we are complaining of pain, feeling angry, or hallucinating. There are predictive psychological tests of memory for conditions in which forgetfulness is the complaint. We could measure the maximum oxygen utilization on a stationary bike of the person who feels easily fatigued with mild exertion. But the brain scan picture, the score on the memory test, and the readout of oxygen consumption are still some steps removed from the personal experience of the pain shooting down the leg, the distress at misplacing the car keys again, or the exhaustion of reaching endurance's limit in a race. Thus, the tension continues between what we can directly measure, in terms observers agree about, and the less accessible feelings and inner experiences of the patient. This tension is one of the sources of the disagreements about what Lyme disease is.

The Science and Art of Medical Diagnosis

Diagnosis is often compared to the detective's process in solving a crime. The physician or police officer draws up a list of "suspects" for the illness or crime. Just as the officer may bring in for questioning people who by virtue of their records may be suspected of having committed the crime, so the physician produces what is known as a "differential diagnosis," a list of specific diseases, in order of probability, to be further considered. This list may be short—as short as one item—if the diagnosis is clear, or it may be long, if the nature of the illness is obscure. The list is refined, resorted, and reevaluated as additional data come in. The evidence for the detective may be in the form of information gained from interviews— personal accounts—or in the form of data from forensic examinations, such as fingerprints, gunpowder residues on clothes, tire tread marks, or DNA sequence profiles. These two forms of evidence are analogous to the symptoms and signs that a physician deals with.

To gather information about the patient and the illness, physicians generally take the following steps: they first take a medical history, next perform a physical examination, and finally order any radiology procedures or laboratory tests that may be appropriate.

Taking a history, as the name implies, involves developing an understanding of the sequence of events that led the person to visit the practitioner's office. The patient tells the physician, either spontaneously or with prompting by questions, about the symptoms she or he is experiencing, when they started, whether they have gotten worse, whether they are continuous or intermittent, what makes them worse, and what, if anything, relieves them. If the physician does not know the patient, he or she will ask about other conditions the patient has now or has had in the past, as well as allergies or sensitivities to medicine, and any diseases that may run in the family.

In an ideal encounter between physician and patient, there is enough time for a thorough history to be taken. Unfortunately, this is less and less often the circumstance, as the time allowed on a cost basis for taking a history decreases. So often times, the "database" on which the physician must form a diagnosis is incomplete. Accordingly, it is helpful when the patient brings in concise written notes about what happened and about the present symptoms. This information is likely to be more accurate than information carried in one's head, and it makes it less likely that anything of importance will be overlooked. Such notes should include the names and dosages of any medicines presently being taken. If a person is seeing a new physician, it is useful to bring a copy of previous medical records that may be pertinent or, preferably, arrange for their transfer ahead of time to the clinic and to the doctor's attention. This can save time and reduce repetition of questions and laboratory tests.

The next step is the physical examination, which can range in extent from limited to comprehensive, depending on the complexity of the illness and whether the physician has seen the patient previously—and if so, how long ago. Usually the physician asks additional questions while performing the exam. During the exam, too, the patient may mention other symptoms that seem unrelated to the present problem but may nonetheless turn out to be important.

After completing the history and physical, the physician often has a pretty good idea of what is going on. If the diagnosis is not clear, at least many diagnoses may be eliminated from further consideration. The physician may decide at this point that no laboratory tests or x-rays are needed and will dispense either appropriate treatment or reassurance—or both. Sometimes, though, the recommendation will be that the patient have some diagnostic tests: an x-ray of some kind; another procedure, such as an electrocardiogram; or laboratory examinations of the blood, urine, or other specimen, like the sputum if the patient has a cough.

These tests may be ordered to confirm a diagnosis that seems likely, given the information provided by the history and physical examination. Less commonly, they may be ordered because it is not at all clear what is going on. In this latter case, the physician depends on the test results to further inform the decision-making process. An example of this situation is a culture of the blood for microorganisms in a case of fever lasting for weeks. In such cases, the physician may suspect that the patient has an infection, say, because of a persistent fever, but may have little evidence to go on in determining what and where the infection is.

As may be obvious by now, a diagnosis—what is written down in the medical chart or hospital record—comes in different degrees of certainty. Some diagnoses, typically those at the beginning of the workup, are provisional. They may be changed when conflicting information is obtained or after what would seem to be an appropriate course of treatment fails. In other cases—and this may occur either sooner or later—the diagnosis is definite.

A definite diagnosis may be arrived at in several ways. Sometimes the story that the patient offers or that the physician elicits is so typical of a particular condition that further examinations are only confirmatory. An example of this would be a history of migraine headaches. The physician can make a diagnosis at a high level of certainty just from hearing about the patient's headaches and the pattern of headaches. The profile of symptoms is consistent with migraines. In other cases, the disease may be suspected from the patient's story, but the physical exam supplies the proof. Examples of this are asthma, confirmed when the physician listens to the chest and hears wheezes, and many diseases of the skin, which is an "organ" that is accessible to the eyes and touch of the physician.

Current economics of health care in the United States encourage use and dependence on laboratory tests and other diagnostic procedures. These conditions include time pressures, which make it difficult for the physician to spend the appropriate amount of time talking to the patient, and the perceived threat of medical malpractice suits. The legal threat may cause the physician to order tests to rule out even diseases that are relatively unlikely in light of the medical history and physical examination. This behavior is rationalized on the grounds that if a serious disease is missed, the physician may be sued, even if it can be agreed that the disease was very unlikely to begin with.

Physicians who are faced with a patient's bewildering collection of symptoms are in some ways relieved when a "disease" is discovered through a laboratory or an x-ray procedure. They can now turn to the impressive armamentarium of medicines or surgeries available to deal with most problems. In achieving the goal of specific disease diagnosis, the physician has successfully followed the lead of his or her mentors. But there is an inherent danger in following the model too strictly: the "disease" may be managed by the book (in this case, the medical textbook or the increasingly utilized guidelines published by professional societies) regardless of what the patient is actually experiencing.

More and more people are recognizing that illness and the perception of illness are affected by many factors besides the malfunctioning organ. These include cultural and personal issues surrounding what an illness means to a patient. People express their diseases in different ways. In one patient, a "disease" may be diagnosed and the correct treatment given, but the patient still feels ill. When the same "disease" is diagnosed in another patient, and the same treatment is given, the outcome may be very different.

Figuring the Odds

One of the enduring aphorisms of medicine (in North America, at least) is this: if you hear hoofbeats outside the house, think of horses and not zebras. Indeed, to denigrate a diagnosis made by a student, trainee, or colleague, one needs only to call it a "zebra," meaning that the diagnosis is very unlikely in that person in that locale. For example, a diagnosis of malaria in a person with a fever who has never been outside of Minnesota is a zebra. It is remotely possible that the person has malaria, perhaps contracted from a blood transfusion, but it would be inappropriate to inflate the odds for a diagnosis of malaria, particularly if doing so meant that a more likely and more treatable diagnosis was overlooked. If all the "remotely possible" diagnoses for every illness were explored or treated, the hospital bill would be huge and the patient overwhelmed with tests and medicines.

Lyme disease, like malaria, does not occur everywhere in the world. Nor, as discussed in chapter 4, is it found in every part of the United States. Even within states that report many cases of the disease, the risk of contracting it varies greatly by locale. In many places, a person's chances of getting Lyme disease are zero. Essentially the only way that a person can get the disease is by living in or visiting a place where ticks with the spirochetes exist. For example, although Westchester County, New York, reports many cases of the infection each year, a resident of that county who stays in strictly urban areas, where deer are not present, has little or no risk of Lyme disease.

What all of this is leading up to is a recognition of the importance of assessing risk. Estimating the individual's risk of getting Lyme disease is an important part of diagnosing the disease. Estimating risk is much like estimating the odds of winning a lottery. If you have a thousand tickets for the next drawing, the probability that you are holding a winning ticket is higher than if you had only one ticket. If the patient has been living in or has visited an area in which the infection is known to occur, then he or she is holding a fair number of Lyme disease "tickets." Additional "tickets" are obtained by spending time outdoors in areas where deer have been observed. When a diagnosis of Lyme disease is being considered in a person living or working in an area where the infection has not been confirmed, then that person's travel history becomes important.

A caution: complicating risk assessment based on location is the fact that Lyme disease ticks and *B. burgdorferi* spirochetes are spreading to areas, such as Vermont, northern Maine, Michigan, Canada, and Virginia, where they were not detected when Lyme disease was first described. Whether Lyme disease continues with its expansion remains to be seen, but it is likely, as chapter 4 discusses. This is not a rapid spread, such as what we see with a new strain of the influenza virus, but it means that risk maps will need to be updated periodically. Residents in Lyme disease–free areas bordering the current geographic limits of the disease cannot be complacent. Another consideration is other diseases that are carried by *Ixodes* ticks (chapter 11). These other pathogens may have somewhat different patterns of distribution and rates of spread than *B. burgdorferi*.

Listening to the Symptoms

Some symptoms and signs are either highly uncommon or nonexistent for Lyme disease. If these symptoms occur, another diagnosis should be considered instead of Lyme disease, or at least in addition to it. These include, but are not limited to, the following: bone-shaking chills; sweating at night that drenches the bedclothes; diarrhea with several bowel movements a day; abdominal cramps; painful urination; a persistent cough; heartburn; nasal discharge; epileptic seizures; spitting up blood; edema of the legs; vaginal discharge; painful, swollen finger joints alone; marked weight loss; and severe pain in the chest.

After this litany of what Lyme disease is not, it is helpful to remember that a patient's history can be so suggestive of Lyme disease that the diagnosis could be made just by hearing it. One such history concerns a person who was at risk of infection by virtue of residence and activities and then developed a targetlike skin rash in the late spring and a few weeks later had weakness on the side of his face, a profound slowing of the heartbeat, and eventually recurrent pain and swelling of a large joint. Other than Lyme disease, there are few (if any) diseases that this could be.

Such textbook histories are in reality seldom encountered. The rash may not have been noticed or clearly recalled. Perhaps only one of the complications occurs, and there may be some exceptional features to it. But this idealized case teaches that symptomatic infection with *B. burgdorferi* evolves over time with a certain pattern. The course of the infection can be divided into localized early infection; disseminated early infection; and late infection. These three designations roughly correspond to periods, measured from the start of infection, of days, weeks, and months, respectively. This temporal pattern to the illness can be an important clue that someone has Lyme disease: many patients with *B. burgdorferi* infection do not present with all the classic manifestations of Lyme disease, but a physician can still discern a waxing and waning of an illness lasting weeks to months.

The Physical Exam: The Skin Rash

Putting a name, like erythema migrans, to a skin rash is akin to identifying what bird you are seeing in the binoculars. If you spot a bird that is new to you, you can look in the guidebooks and often identify it on the basis of the drawings or photographs and some of the listed characteristics. But if what you observed is somewhat different from the bird depicted in the guidebook, you have less confidence in the accuracy of your identification. It helps then to have a longtime birder at your side to confirm the sighting. Likewise, a physician in Alaska who sees in July a patient who just returned from New Jersey with a rash may come up with the correct diagnosis of early Lyme disease by looking at some pictures of erythema migrans in textbooks or on websites. But if the rash does not look exactly like one of the pictures, the doctor may be stumped or more cautious. Contrast that with a primary care physician based in New Jersey who may have seen hundreds of examples of erythema migrans over the years and has developed a sense of the range of what the rash may look like and the contexts in which a true Lyme disease rash occurs.[2] Experience—practical knowledge—counts. This should be kept in mind for what follows: an introduction, not the last word.

As with bird identification, it helps to take into account the time of year and the geographic location. A rash appearing in February in New York or Massachusetts is unlikely to be erythema migrans. In most cases of Lyme disease with a rash in the northeastern and north-central United States and adjoining areas of Canada, the rash appears in the late spring, the summer, or the early fall. Lyme disease starts with a tick bite, and ticks are active only during certain times of year. That rule of thumb does not hold in regions, like some parts of California, where *Ixodes* ticks are more active in the cooler, wetter months of winter than in the hot, dry summer. But the point is that the erythema migrans skin rash occurs during or no more than a few weeks after the period when the nymphal ticks are biting. One can get erythema migrans from the bite of an adult tick, which is usually encountered later in the summer or in the fall, but this is not common.

As described in chapter 1, the rash starts at a place where a tick embedded. Occasionally a small tick engorged with blood may still be present. Many people develop a small, slightly elevated area of redness around a tick bite or an embedded tick. This can occur with many types of ticks and in the absence of any spirochetes. In these cases the body is reacting to the tick itself, or to substances the tick left behind. Typically, this limited rash—usually smaller than a quarter—appears within a day or two of the tick bite and then fades over the course of a few days. As discussed previously (chapter 1), some tick bites become infected with other types of bacteria that are already present on the person's skin. The rash and swelling, the cellulitis, from these staph or strep bacteria can spread very rapidly and need immediate medical attention. Some of these infections can become life threatening. Cellulitis may follow other small breaks in the skin besides a tick or insect bite. High fever and a painful rash are more suggestive of cellulitis than of erythema migrans.

The rash of erythema migrans begins like a reaction to the tick bite itself—that is, as a small area of redness. The onset of erythema migrans is later than that of the tick bite reaction, however: it occurs three to fourteen days after the bite instead of one to three days. Erythema migrans is also less likely than bite reactions to itch. In fact, if the rash appears in an area such as the back of the leg, where a person does not usually look, it may be missed entirely (as in Scene 1 above). A tipoff feature of the erythema migrans rash might be its ringlike appearance: a red circle enclosing a paler center. As described above, there may be another ring within, giving the rash the appearance of a target. Many Lyme disease rashes are solidly red with no or only a hint of clearing in the center, but the outer margin has a distinct, not blurry, edge. And erythema migrans rashes are not always circular: triangular or almond-shaped rashes can occur during the early infection.

What the early Lyme disease rash of erythema migrans usually is not is scaly, crusty, or blistery. Hives or welts are not typical and suggest another diagnosis, such as an allergic reaction to an insect bite or sting. A rash on the soles of the feet or the palms of the hands is very unusual in Lyme disease. The combination of a skin rash and blisters or ulcers of the mouth is probably not Lyme disease. Neither is a head skin rash with loss of hair.

A common misconception about the rash of Lyme disease is that it can commonly return during late disease, months to years after the onset of the disease. Many people wonder if they have Lyme disease because they have heard or read that people with the disorder have the symptoms they are experiencing: a rash and joint pains, perhaps with fatigue, headache, and other symptoms as well. During the early phase of Lyme disease, people may have the rash and feel achy all over. They may even experience a return of the rash with multiple red patches in the first weeks of the infection. This is the result of the spirochetes spreading, through the blood, to other areas of the skin. But the combination of either joint pains and a rash or extreme fatigue and a rash would not be expected to occur months or years later. If people with true Lyme arthritis do get a rash later in the disease, it is more likely for another reason, such as an allergic reaction to a medicine.

The only skin rash from Lyme disease that lasts for months to years is usually known by its tongue-twisting Latin designation, "acrodermatitis chronica atrophicans."[3] The name essentially means a long-lasting rash of the arms and legs that leaves the skin thin and wrinkled. This rash may evolve directly from erythema migrans, or it may emerge after a disease-free interval. The involved skin may be reddened, but overall, the rash resembles an old burn. Often there are effects on the nerves of the same limb with the rash. Almost all cases of acrodermatitis have been reported from Europe and Russia. It is rare, if it occurs at all, in North America, probably because only a type of the *Borrelia* spirochete that occurs on the Eurasian continent can cause this form of Lyme disease.

The other skin condition that is suggestive of infection with a Lyme disease spirochete is "lymphocytoma." Like acrodermatitis, it is much more common in Europe than in North America. Lymphocytoma is a deep reddening and hardening of one of the ear lobes or of the areola around a nipple.

The Physical Exam: The Rest of the Body

The remainder of the physical exam may reveal other signs of infection with *B. burgdorferi*, but these signs are not as suggestive for Lyme disease on their own as the rashes erythema migrans or lymphocytoma.[4] There is nothing unique to Lyme disease about a weakness in facial muscles, swelling and pain of a joint, or a slowing of the heartbeat. For each of these disorders, there may be other explanations. Of course, if all three conditions occur at the same time or in succession, then Lyme disease is the number one suspect, but such a case seldom appears.

The examining physician may discover other abnormalities during the physical exam. In the early stages of the disease, when a rash is usually present, these include a redness at the back of the throat; swelling of some lymph nodes of the neck, armpits, or groin; and redness of the eyes. If there is a fever in Lyme disease, it is seldom higher than 102°F (38.9°C). A temperature higher than this is probably caused by another infection or condition. The pulse may be more rapid than normal, especially if there is a mild fever. If the pulse is much slower than usual, the heart may be involved. Blood pressure seldom is much higher or lower than usual for the patient.

When the patient or his or her family expresses concerns about forgetfulness, inability to concentrate, sleep difficulties, or irritability, the patient's "mental status" is checked. At that point in the examination, the physician asks the patient to remember objects and to perform some mental and written tasks. Such tests commonly demonstrate that concerns about the patient's supposedly failing abilities are groundless; as a consequence, these tests are often reassuring. In other cases, real impairments are revealed that may indicate brain involvement. Again, there may be other explanations besides *B. burgdorferi* infection for this. Because major depression can produce the same symptoms and mental deficits, physicians may also assess patients' self-esteem and ask patients whether they have been feeling inadequate or despondent, have noticed any loss of interest in life's pleasures, or have had any thoughts of suicide.

Do-It-Yourself Diagnosis?

Lyme disease is seldom if ever discovered during a routine physical examination. If a person has been feeling well and goes to the physician for a checkup, the physician might detect high blood pressure, a breast lump, or an enlarged prostate; but the physician is unlikely to "find" Lyme disease during such an examination.

Usually, the first step toward a diagnosis of Lyme disease is taken by people who conclude that something is wrong with them or with their child. Such people might then seek out a medical expert to identify exactly what the problem is. But we diagnose for ourselves many common illnesses and disorders, such as a head cold, a sprained back, or a temporary headache, and we decide how to treat the problem as well. Even when we do seek out a practitioner, we usually have localized the disorder to a specific part of the body. Many times we suspect that we know what the diagnosis will be, especially if we've had the same or a similar illness in the past. Sometimes we seek out a physician because we are afraid that we have something seriously wrong with us, such as cancer.

How do we conclude that the pain in our stomach is cancer, or that the pressure in our chest after climbing the stairs is heart disease? We experience pain, shortness of breath, nausea, cramps, "pins and needles" sensations, and so on. In the same way, people with *B. burgdorferi* infection experience one or more of various symptoms, but there is nothing in the symptoms themselves that says, "This is Lyme disease." It is only after we have read or heard about Lyme disease that a particular symptom or particular group of symptoms suggest to us that diagnosis. This tendency to attribute our bodily dysfunctions to a specific disease—this tendency to diagnose ourselves—is a double-edged sword.

The benefit of self-diagnosis is that a disease that is treatable is appropriately paid attention to, perhaps at a point when the symptoms themselves are not particularly disabling. A woman feels a pea-sized mass in her breast and justifiably wonders if this is cancer. This prompts a call to her doctor, and a mammogram is performed. A man has transient pain in his chest and left arm and rightly wonders whether he is at risk of a heart attack. A treadmill stress test later that week reveals poor blood flow in his heart's arteries. A Connecticut resident with knee pain and swelling four months after an episode of skin rash wonders whether this could be the same tick disease that his neighbor had last year. So he asks his physician whether this could be Lyme disease, and the physician agrees that it might be after a history and a physical and orders a blood test. It is positive, and antibiotics are started that week. A person who suspects that he or she has Lyme disease may even go directly to someone who has a reputation, by word of mouth, websites, or social media, of specializing in diagnosing and treating this disorder.

A downside of self-diagnosis, for health care professionals as well, is the tendency to force the symptoms to fit one disease rather than dealing with the symptoms as they are and trying to find the most plausible diagnosis among the many possibilities. Once the idea of Lyme disease is raised as an explanation for a person's physical problems, then it is tempting to fit all the symptoms into this diagnostic bag. It may be human nature to seek an explanation for the bad things that happen to us. For many people, any diagnosis, even an inaccurate one, seems to be preferable to uncertainty, at least in the short term. Physicians can be equally ill at ease without a specific diagnosis to act on. In some situations (like Scene 2 above), the patient's proposal of the diagnosis of Lyme disease is accepted by the physician, and antibiotic therapy follows. This is fine if the person really does have an infection with *B. burgdorferi*. But what if the person does not?

One outcome of such a diagnostic error is the phenomenon of nondisease. This is the identification of disease when it is not actually present. Examples of nondisease are the labeling of a normal variation of mood as major depression, the restlessness of a young child in the classroom as attention deficit hyperactivity disorder (ADHD), or the diagnosis of a heart disorder in a teenager who has an inconsequential heart murmur. A person or that person's physician has a checklist of the possible symptoms of Lyme disease and concludes that occasional joint aches, fatigue, and headache are Lyme disease instead of the effects of a stressful lifestyle that they could be.

One consequence of a nondisease diagnosis is further testing and investigations; these have their costs in terms of money, test complications, and mental anguish. In a case of "non–Lyme disease," there may also be unwanted side effects from antibiotic therapy (chapter 9). An insidious effect of the nondisease phenomenon occurs when someone who is actually in good health assumes the role of a sick person. A child erroneously diagnosed with heart disease or a food allergy may be needlessly restricted from physical play or from eating certain healthful foods.

There is an even more unfortunate outcome of a false diagnosis of Lyme disease: when a false diagnosis is settled on, diagnosis of another disease, such as cancer, depression, or multiple sclerosis, may be overlooked or delayed. In this case, not only has the wrong diagnosis been made—with the repercussions of that error—but patients miss the opportunity to receive proper treatment for what really ails them.

One way to minimize incorrect diagnoses of Lyme disease is for the physician to listen to the patient's story and to place the patient's illness into the context of the risk of exposure to disease-bearing ticks. This may mean confronting the doctor—not an easy thing for a patient to do—about taking the time to listen and not rush through a set of questions. Certainly, a checklist approach with inordinate emphasis placed on nonspecific symptoms such as fatigue should be avoided. The temporal sequence of events is also important in the history: the arthritis should follow the skin rash and not the other way around.

Back to the Top and the Two Scenes

In Scene 1 there is a good justification for a clinical diagnosis of early Lyme disease on the basis of the history and the physical exam. As we will see, carrying out a laboratory test or some other procedure is not likely to change that conclusion. So treatment with an appropriate antibiotic is started. The situation in Scene 2 is quite different. There is little justification for a clinical diagnosis of Lyme disease on grounds of the risk of infection, the history of the illness, and the physical examination in this case. If Lyme disease is part of the differential diagnosis, it is probably low down on the list. But for other reasons, the decision would be the same as in Scene 1: a Lyme disease test is not indicated. Between these two polar opposite situations are others where appropriate and prudent application of laboratory tests reasonably supplements a thoughtful and skillfully performed patient history and physical examination. Blood tests and other diagnostic procedures are considered in the next two chapters.

Chapter 6

Laboratory Tests
The Basics

It is not uncommon to see patients who think they have been thoroughly studied and have obtained an undoubtedly correct opinion because they have had an endless number of tests made and have been furnished with extensive reports, often meaningless. In turn, the physician at times seems to feel that "if all the tests are done" then he had overlooked nothing; such a man often gives little attention to the accuracy of the tests and loses sight of the fact that proper interpretation of the results is fundamental.

Dr. George Minot, 1922

This remark could have been made today.[1]* In fact, it was made almost a century ago, by Dr. George Minot, a future Nobel laureate, at a meeting of the Massachusetts Medical Society. We may forget that many of the laboratory tests and other procedures we are familiar with today were available to physicians many decades ago. These include the blood cell count, urine analysis, electrocardiogram, x-rays, cultivation of bacteria (but not yet viruses), and even tests for antibodies to certain pathogens (most importantly then, syphilis). Although some of the tests that Dr. Minot was concerned about were eventually discredited or supplanted by improved versions, contemporary medicine probably also has its share of tests that will eventually prove to be of little utility for good care. At the same time, there are so many more tests to choose from than in

---

*Excepting the dated assumption about the gender of the physician.

Dr. Minot's day. This is especially true for laboratory tests looking for antibodies to pathogenic microbes, only a handful of which had been identified by 1922. And there are entirely new technologies, such as CT (computed tomography) and MRI (magnetic resonance imaging) scans for imaging various body parts as well as the polymerase chain reaction (PCR) for detecting very small quantities of an organism without cultivation. Antibody tests and PCR assays are particularly relevant for Lyme disease diagnosis, so I will focus on those.

The previous chapter covered physicians' use of a patient history and physical examination to arrive at either a single diagnosis, which leads to a treatment plan then and there, or a list of possible diagnoses in rough order of probability. In the latter case, the doctor may still proceed, even under uncertainty, and write a prescription with the top-ranked diagnosis in mind. But in most cases, the uncertainty will prompt further studies, which in effect enhance the capacities of the physician to detect the signs of disease. In this sense, an x-ray extends the visual sense to "see" inside a body, and a blood test more fancifully extends the sense of smell to be able to detect various different substances in the blood.

So, a "clinical diagnosis"—that is, one built on the history and the physical, before any laboratory test or radiology procedure is performed—may or may not be enough. In the hypothetical example of a patient with the full set of risk factors, symptoms, and physical signs of Lyme disease, a clinical diagnosis alone would be on firm ground and would reasonably result in treatment with an antibiotic. But in another case, a physician may not be ready to treat on strictly clinical grounds, especially when there is no rash or a rash passed without ever being observed by a doctor. If there is reason to think that the patient has been infected long enough for the immune system to be responding (chapter 1), then a laboratory test for antibodies to *Borrelia burgdorferi* is usually justified. A positive test result advances the diagnosis from a clinical one to an "etiologic diagnosis" of *B. burgdorferi* infection. The actual cause, or "etiology," of the disorder is identified. (Etio-, or aetio-, derives from the Greek word *aittia* for "cause.") The etiology of Lyme disease in North America is *B. burgdorferi*. In Europe it could also be *B. afzelii* or *B. garinii*. The etiology of AIDS is the human immunodeficiency virus, or HIV. The etiology of malaria is one of three species of the *Plasmodium* parasite. And so on.

Screening for Disease

In the diagnostic process summarized above, the decision to order a laboratory test followed the history and the physical. It did not precede it. The cart was after the horse, not before it. The aim of the test was to help the physician resolve between two or three (maybe four) different possible diagnoses, one of which was Lyme disease. In the doctor's mind, Lyme disease was highly plausible but not a sure thing. The doctor may not consciously have thought about it this way, but if you asked her or him, "Do you think this patient before you is more likely to have Lyme disease than the person in the street outside the office?"—the doctor would likely answer, "Yes, of course."

But what if the patient walks into the office with a lab result in hand reporting a positive test for Lyme disease? Let's say that the blood test was performed as part of a community-wide survey of residents, and the person was informed of a positive test. When blood is tested for a disease without a good reason to suspect the infection's presence, beyond the person tested being a member of an at-risk group, it is said that the blood is being "screened" for disease. As the word "screening" implies, this is like putting a lot of gravel from a stream bed onto wire mesh and looking for the rare gold nugget.

The cholesterol test performed in pharmacies and clinics is an example of a screening test. This test is given to everyone, even people without symptoms, and is intended to identify people with high cholesterol levels who may not have any other indication that they are at high risk of a heart attack or stroke. Another example of a blood screening test is the infamous blood test that couples once routinely had to undergo before getting married in most states. The rationale for this prenuptial check for evidence of syphilis was that this is an infection that can be silent for many years and that can be passed on to one's children; if everyone was screened, the testing process would discover those who have the disease and are therefore in need of treatment. Discovery of an undiagnosed case of syphilis in this situation was about as infrequent as discovery of a gold nugget in a slurry screen, but screening was rationalized in these situations for public health reasons.

Can a blood test be used to screen for Lyme disease? In many instances, the test has served this aim in practice if not in conscious intent. Consider the person who has some symptoms that suggest Lyme disease but could also be the symptoms of several other conditions, and in this instance, the patient has been at little risk of acquiring the infection (as in Scene 2 in chapter 5). The physician thinks the chances of the person having the infection are low, yet he or she orders the test anyway. The physician's reasoning may have been something like this: "If the test results are positive, then I have the answer, and treatment can be started. But if the results are negative, then at least I made sure that Lyme disease was ruled out as a consideration. Lyme disease is something I don't want to miss, because it's treatable."

On the surface this sounds reasonable, but it is also what Dr. Minot was getting at in the epigraph. The unexpected finding of Lyme disease in someone with atypical symptoms can be gratifying for both patient and physician. But are the tests so accurate that we can accept the laboratory results as proof of Lyme disease even when the findings from the history and the physical examination argue against this diagnosis? Answering these and other questions will take us on a tour of the various tests for Lyme disease. We'll see how they are done and how reliable they are. We turn first to the most commonly performed blood test for evidence of Lyme disease, the detection of antibodies in the blood. This will serve as a model for understanding the other laboratory tests described here.

Looking for Antibodies in the Blood

The laboratory may examine your blood or tissue for evidence of the spirochete itself (these tests, called "direct tests," are considered later in this chapter). Much more commonly, however, the laboratory looks for indirect evidence of the infection—namely, the presence in the blood of antibodies to *B. burgdorferi*. Antibodies are the protein molecules that mammals and other vertebrates produce as a defense against infection (see chapter 1), and in a sense, they are tailor made for the invading microorganism. That is, the antibodies formed in response to an infection are usually unique for that infection.

The antibody test is not foolproof. For one thing, when the laboratory technician finds antibodies to *B. burgdorferi* in the blood (or antibodies to the influenza virus, or to any other infection), this does not necessarily mean that Lyme disease (or influenza) is still active. Some types of antibodies may be present months to years after the infection is over. Another shortcoming of an antibody test is that the antibodies the laboratory finds may not actually be from a Lyme disease infection. They may be the byproduct of another type of infection or type of disease.

When a Lyme disease test is positive but the person does not really have *B. burgdorferi* infection (or ever did), the reaction is called a false-positive. A false-positive test may occur when the blood contains antibodies to a spirochete that resembles but is not identical to *B. burgdorferi*. For example, the antibodies may have been formed in response to a related *Borrelia* species, like *Borrelia miyamotoi* (chapter 11), rather than to the *B. burgdorferi* itself.[2] The two *Borrelia* species have enough in common that an antibody originally formed in response to infection by one will also react with the other, even if it has not caused any disease in the patient. A false-positive test may also result when people have antibodies that are particularly "sticky." (What is behind these and other misleading reactions will be described in more detail along with the specific tests.)

Even with its deficiencies, the antibody test, performed by a skilled technician and in an approved laboratory, can provide useful information. Interpreted with realistic expectations, the test can make a difference in a diagnostic decision. If you were going to have only one test for Lyme disease, the antibody test would be the best choice at the present time. To better appreciate its strengths and weaknesses, let us follow what happens to a blood sample in the process of testing for antibodies to *B. burgdorferi*.

The EIA test

EIA stands for enzyme immunoassay. The "immuno-" and "-assay" parts of the name indicate what it's for, and we will consider the "enzyme" part later. The test is also called an ELISA or Elisa (pronounced "ee-lie-sah"), for enzyme-linked immunosorbent assay. For the EIA, a tablespoon or two of blood is drawn into a glass tube with a red stopper. In the "red top tube," as it is called, the blood clots within a few minutes. The blood cells congeal into a dark red mass, and the straw-colored serum can now be seen. The laboratory technician separates the clotted red blood cells from the serum by spinning the tube rapidly in a centrifuge. The clot goes to the bottom of the tube, and the liquid serum remains on top. The serum is removed, and the red blood cell clot is discarded after sterilization. (The test can also be performed on plasma, which is obtained from blood first treated to prevent clotting and then removed of cells by centrifugation.)

The procedure for doing a Lyme disease EIA test is fairly standard across different platforms and operators. First, the serum is diluted with a solution without antibodies; a common ratio is 1 part serum to 99 parts of the reagent solution. If antibodies to *B. burgdorferi* are present, they can usually still be detected even with a concentration of the patient's serum at only 1 percent of what it was. If the serum is not diluted to that extent, false-positive reactions are more likely. Antibodies stick nonspecifically to the assay materials when in high concentration, and this effect needs to minimized. On the other hand, if the serum is diluted too much, the assay may be false-negative; that is, there is a readout of "negative" when the patient truly has Lyme disease. The antibodies are there but in too low a concentration after the dilution for the test to register as "positive" by the usual criteria. You can see that the result might tip either way depending on what dilution is chosen. Preferably, several different dilutions are tested, as was done in the early days of antibody testing, but this is seldom done anymore.

The diluted serum is then put in what looks like a miniature plastic version of a muffin tin. Stuck to the bottom of each depression, or "well," is a small amount of the Lyme disease spirochete or isolated proteins.

When whole cells are used, they are dead and, in most tests, broken up into small pieces. If there are antibodies in the serum or plasma to parts of the spirochetes or the isolated protein, these will bind to the bottom of the well and will not be washed away when the wells are washed with a saltwater solution. These bound antibodies can be detected by adding to the wells a second solution, which contains antibodies to antibodies. In other words, the presence of human antibodies stuck to the spirochete parts is revealed by using a special antibody that reacts with, or recognizes, human antibodies. This antibody is generally produced in a goat, sheep, or rabbit by vaccinating these animals with human antibodies. Just as people and animals are immunized with a vaccine to provide them with a ready supply of antibodies against this or that infectious agent, these animals are vaccinated with human antibodies, and therefore they produce antibodies that react to human antibodies.

If the second antibodies, that is, the antibodies to human antibodies, were used just as they came from the animals, it would be difficult to tell whether they were present on the bottom of the pan because antibodies are colorless. The trick used to solve this problem is to link another protein, an enzyme, to the second antibody. (This is where the "E" in EIA comes from.) This enzyme, like other enzymes, works by changing one chemical compound to another. In this case, the chemical molecule changed is a dye. And when the dye is altered by the enzyme, its color changes. In the most frequently used reactions, the chemical compound is changed from white to yellow or blue. Each enzyme molecule can act many times over to change the compounds, making it possible to detect even small amounts of antibodies in the serum.

So the final steps in the test method itself are these: The solution containing dilute second antibodies with their attached enzyme is added to the wells. After an hour or so of incubation, the second antibodies not bound to the bottom of the well are washed away. The presence of the bound second antibodies is then revealed by adding the chemical that turns yellow or blue. The change of color is visible to the eye, but the degree of color change is routinely measured on an instrument that can provide a value for the intensity of the color of the chemical dye. In general, the higher the concentration of antibody in the specimen, the greater the intensity of the dye's color and, thus, the higher the measurement value on the instrument.

The raw results of an EIA test are usually a ratio derived by comparing the instrument value obtained with the specific patient's serum to the value obtained with sera from people without Lyme disease. For example, let's say that the intensity of the color change in the serum of someone with suspected Lyme disease at a dilution of 1:100 was measured on the instrument as 0.74 out of a possible 1.50 or so. On the same day, the serum of someone without Lyme disease (termed a "negative control") produced a value of 0.20. Accordingly, the ratio, or "index" as it often called, is reported as 3.7 (0.74/0.20).

Advances in the EIA

One way to improve the performance of the EIA test is to eliminate proteins that are either cross-reactive or irrelevant. Without the cross-reactive proteins, the test could be more specific, since there would be less chance of an adventitious reaction. By reducing the number of irrelevant proteins, the concentrations of the proteins that actually count could be increased in the EIA test well. This could increase the sensitivity of the assay. These improvements could be achieved either by removing the cross-reactive and irrelevant parts from test material or by mixing together as a "cocktail" those proteins that are most desirable for diagnosis. The second approach is easier than the first, because it can make use of recombinant DNA technology to "clone" the genes for the proteins of interest. Once those clones have been obtained, they can be engineered to produce the protein in another, harmless bacteria in the laboratory. Once separated and purified, the recombinant proteins can be used singly or in combination in an EIA or some other format.

The first substance to be exploited in this way goes by the shorthand name of VlsE protein.* This protein and its family members are unique to *Borrelia* spirochetes. It does not occur in other bacteria, viruses, or par-

---

* I participated in the discovery and characterization of the VlsE protein and am an inventor on granted patents that are owned by the University of Texas and nonexclusively licensed to several diagnostics companies for commercial assays.

asites. Either the whole protein or one part of the protein, known as the "C6 peptide," can be used by itself in place of the whole cells of *B. burgdorferi* in the EIA test. The sensitivity and specificity of the EIA test based on the VlsE protein or the peptide appear to be at least as good as the whole-cell EIA in terms of sensitivity and specificity.[3]

A more recent EIA test, which has obtained FDA approval and is based on recombinant proteins, contains not only VlsE but two other proteins that are known to be specific for *Borrelia* species and to commonly elicit antibody responses during infection. According to publicly available documents submitted to the FDA as part of the approval application, this EIA test performs at least as well as tests based on whole cells. (The next chapter describes the approval process of diagnostic tests in more detail.)

These recombinant protein-based versions of the EIA have the advantage of restricting the reactivities to well-defined proteins in pure form. But there may be something lost by abandoning tests based on whole cells of the bacteria. Some substances that elicit antibody responses during infection are not proteins.[4] These substances are called "glycolipids" and are combinations of a sugar (glyco-) and a simple fat (-lipid). Glycolipids cannot be cloned the way proteins can be, because their synthesis is more complex, involving several genes instead of just one.

EIA Test Interpretation and Its Pitfalls

At this stage, an interpretation of the raw test result has not been made. What usually happens is that the laboratory reports whether the index value is positive or negative. The laboratory thereby reduces what is actually a point on a wide scale (a quantitative result) to a yes or no answer (a qualitative result). Sometimes a result in a gray zone between positive and negative is called "borderline," "indeterminate," or "equivocal" by the laboratory.

If a value on a continuous scale is finally reported as negative or positive, it means that someone, or more likely a committee, has specified a "cutoff point," a test value at or above which the test result is considered to be positive and below which the test result is considered to be negative. How is the cutoff point set? Selecting a cutoff point would be simple if people without Lyme disease showed no reactivity in the assay even with the least diluted samples of serum. But this is not what has been found. On the contrary, some healthy people, the negative controls, have had detectable antibodies that bound to spirochete parts in the well of the test plate. These people seldom had titers or color values that were as high as those of Lyme disease patients, but the two groups did overlap. There was no value below which all control sera fell and above which all Lyme disease sera fell.

How could this be? Members of a control group are supposedly free of other diseases. There are three possibilities. The first is that some people were infected (or had been infected) with another microbe that was similar enough to the *B. burgdorferi* microbe for some of the antibodies to bind to proteins of both microbes, thus confounding the assay. This could be syphilis, which is caused by another type of spirochete. It could be infection with one of another group of *Borrelia* species, those that cause relapsing fever, including *B. miyamotoi*, which is carried by the same ticks that transmit *B. burgdorferi* (chapters 3 and 11). Or, the infection could be as innocuous as some mild gum disease. Spirochetes that are not too distantly related to the Lyme disease agent occupy the crevices between teeth and gums. If their numbers get very high, people may develop antibodies to these mouth spirochetes. Some of these antibodies may react with parts of *B. burgdorferi*, and as a consequence, wells with these sera show color changes, enough perhaps for a false-positive reaction.

The second possible cause of a false-positive reaction among controls is the presence, in the blood, of antibodies that—as mentioned above— are unusually sticky. There are millions of types of antibodies, and some of these have the property of adhering to a variety of surfaces, including the spirochete parts and even the plastic at the bottom of the test well. These sticky antibodies may occur in otherwise normal people, but they are more common in individuals with such autoimmune diseases* as lupus erythematosus (often called lupus, or SLE) and rheumatoid arthritis. These sorts of false-positive reactions can also occur with other in-

---

* As the name suggests (the prefix auto- means "self"), in these disorders people have an immune response against their own tissues.

fectious disorders in which there is a great outpouring of antibodies. The immune system is stimulated not only to make antibodies against infecting pathogen but also to produce willy-nilly other antibodies, some of which may bind to the substances and materials of the *B. burgdorferi* EIA test. This happens in some cases of infectious mononucleosis, which is caused by a virus, and in malaria, which is caused by parasites.

The third possibility is that some members of the control group actually were infected at some time with *B. burgdorferi*. When there is no rash, and when the early infection is not obviously followed by arthritis, a nerve disorder, or involvement of the heart, there is nothing to suggest Lyme disease. Infected people may think that they have the flu or a summer virus. The disease, if any was noted at all, passes, while antibodies to the Lyme disease spirochete appear in the blood in response to the mild infection. These antibodies may persist in the blood for months or years. The chance that one of the apparently uninfected controls has had a prior *B. burgdorferi* infection is obviously greater in areas where Lyme disease regularly occurs. That is why negative control sera are best obtained from places far from a Lyme disease area. The only way that residents of such a place can develop *B. burgdorferi* antibodies is by traveling to an area in which Lyme disease transmission occurs. A typical false-positive rate for an EIA test with sera from individuals living in areas where Lyme disease occurs, but who do not report having had the infection, is about 3 percent. The false-positive rate is about 1 percent for sera from people living outside these areas.

Of course, there can also be false-negative results: the patient had *B. burgdorferi* infection, but the EIA test was negative. This might be attributable to taking the blood sample during early infection, before the antibodies had risen to a high enough level for detection (chapter 1). In someone who clearly had erythema migrans in the past (more than one or two months ago), antibiotic therapy in the beginning stage of the infection may nip in the bud the spirochete's stimulus to the immune system. Or, as discussed below, the species or strain of *Borrelia* used for the assay may not have been the best match for what the patient was actually infected with.

The Western Blot Test

Faced with the trade-off between sensitivity and specificity, most would err on the side of reduced specificity if that trade-off could achieve high sensitivity. The reasoning would be, "I'd rather give an antibiotic to someone who may not have needed it than let an infected person go untreated." However, that's an unsatisfactory compromise. Can't we have our cake and eat it, too? The obvious solution would be to improve the specificity of the EIA test (or something equivalent) without sacrificing sensitivity. Until that is achieved and put in practice, the alternative strategy has been this: first use a sensitive but not so specific test and then submit the samples with a positive result on the first test to a second test that has a better performance by the specificity criterion.

For the past thirty years, the second test of what has been an effective scheme has been the "Western blot" assay, also known as the "immunoblot" assay.[5] A Western blot assay involves more work than an EIA test, first to produce it and then to carry out the procedure and analyze the result. Consequently, it costs more to do and is appropriately limited in its application to this auxiliary role.

The Western blot assay is done in several steps. The first involves separating the various proteins of disrupted *B. burgdorferi* cells according to size. This is done in a thin gelatin-like sheet under the influence of a direct electrical current. Essentially, the smaller proteins move more quickly through the nooks and crannies of the gel than do larger proteins. Once this separation has been achieved, the proteins are blotted in an exact replica to another sheet laid on top of the first. This second sheet can be a refined form of paper or a tougher material, such as nylon. From this point on, the procedure is like the EIA test. That is, the binding of antibodies in the patient's serum to proteins is detected by a second antibody with an enzyme attached. The presence of bound antibodies is revealed by a colored band on the blotted sheet. A modification of the assay uses second antibodies that are tagged with a fluorescent dye, and the intensity of the light-emitting bands are measured with a scanner-like instrument.

The Western blot differs from EIA tests in a fundamental way: the individual proteins are distributed one-dimensionally over a strip of the sheet instead of occupying a single, undifferentiated point at the bottom of an EIA well. This is significant because the laboratory can tell which of the numerous *B. burgdorferi* proteins the patient has antibodies against. In the ELISA test, this is not possible. The Western blot result is reported as the number and identity of the colored bands on the blotted sheet. The Western blot can test for either immunoglobulin G (IgG) or immunoglobulin M (IgM) antibodies.

What is the advantage of knowing exactly what proteins someone has antibodies to? The answer brings us back to the problem of antibodies that bind to something in *B. burgdorferi* but actually arose for reasons other than Lyme disease, such as an infection with another type of spirochete or those nuisance "sticky" antibodies. In the EIA test, these misleading antibodies cannot be distinguished from antibodies that are the true products of an infection with *B. burgdorferi* itself. The Western blot's separation of antibodies allows the laboratory personnel to discriminate which proteins the antibodies are binding to.

There are certain proteins of *B. burgdorferi* that are unusual in the world of bacteria and viruses—there seems to be nothing like them among other causes of infectious diseases of people and animals. If an antibody binds to one of these known proteins in the Western blot assay, it is very unlikely that this occurred because of an infection other than Lyme disease. But there are other proteins of *B. burgdorferi* that are similar to proteins of other bacteria, some of which are only distantly related in evolution. Antibodies to these proteins of *B. burgdorferi* are more suspect, especially in the absence of antibodies to the more unique proteins. In fact, if the only colored or fluorescent bands in the blot are proteins that are common to many bacteria, then the Western blot is interpreted as negative.

The flip-side function of the Western blot is to confirm a positive EIA result by revealing antibodies bound to the unique and specific *B. burgdorferi* proteins. The Western blot for Lyme disease is considered positive when a combination of certain proteins (bands) is present. By the most commonly used criterion, five proteins out of a list of about ten must be present for a positive IgG test. There may be some further significance to counting up the number of proteins that are bound. For instance, there may be more key bands on the blot when the infection is long-standing than if it were more recently under way. But the laboratory's interpretation of the Western blot is usually qualitative, not quantitative. The results are reported as either "positive" or "negative," with an occasional "indeterminate" as the output for results that may merit a repeat testing.

The criteria for interpreting the Western blot results in most laboratories in the United States has remained the same for more than twenty years. The set of ten selected proteins and the requirement for at least five bands follows recommendations from a national conference of Lyme disease researchers, public health officials, and test manufacturers in Dearborn, Michigan, in 1994.[6] These recommendations were adopted by the Centers for Disease Control[7] and the Food and Drug Administration as standards for comparison of new tests under development.

The analysis of the blot strips has been improved by using the technology of scanners that provide a quantification of band intensity to supplement human eyes. And there have been advances in Western blots in Europe that as of this writing are not yet approved for use in the United States. These include adding to the traditional whole-cell-based blot an extra band comprising the recombinant version of the VlsE protein[8] as well as blot materials entirely of purified proteins, which are individually painted across the strips at different locations from top of the blot to the bottom. This would be analogous to the EIA with recombinant proteins, but the proteins would be separated one-dimensionally so it would be clear which of the proteins in this "cocktail" are being bound by antibodies. Also under development is application to Lyme disease testing of a technology that combines the advantages of the EIA, which can in principle provide a numerical value for the reaction, and those of the Western blot, which discriminates between different proteins in their antibody binding. In this procedure, the individual proteins are each coated on small beads of different colors. The reaction takes place in a tube with the beads mixed up, but when it comes time to read the results, the beads can be distinguished in the instrument, so a value for each protein is provided.

Antibody Testing for Early Infection

One reason to repeat the antibody test is when early Lyme disease is suspected and the first test result is negative. During the infection's first month, on average only half of truly infected patients will have a positive antibody test. Although an antibody test is usually superfluous when there is a good history of exposure and an erythema migrans rash, a summer "flu" illness alone in a Lyme disease–risk area may warrant a repeat test two to four weeks after a negative test result. The patient may be asymptomatic by the time of the repeat blood sample, but a positive test the second time around may justify either antibiotics or a close watch for development of symptoms and signs of late infection. In most cases, however, there is little point in repeating the test if antibiotics are prescribed at the first visit. The treatment decision has already been made, and antibiotic treatment at an early enough point in the infection may so reduce the number of spirochetes that a full immune response never develops.

An alternative way to tell in the laboratory if the infection has occurred within the past few weeks is to test the blood for a special type of antibody called an immunoglobulin M, or IgM. IgM antibody is the first type of antibody to be produced during a new infection of any sort. As another antibody, the immunoglobulin G, or IgG, begins to rise in the blood, the IgM antibodies peak and then usually recede over the next several weeks to months, usually to undetectable levels. IgG antibodies join the battle later but generally remain on the scene for months or years. Thus, the presence of IgM antibody to *B. burgdorferi* is evidence that an infection is either going on now or is recent. There are IgM-specific versions of both the EIA and the Western blot test. The IgM EIA is performed as the regular EIA is, but the second antibody binds only to human IgM antibodies. There would also be a different cutoff point for calling a positive result. For the IgM Western blot test, the criteria for a positive test requires fewer bands than for the routine blot, which is based on the binding of IgG antibodies.

One drawback of the IgM EIA and Western blot tests, though, is a higher frequency of false-positive reactions than is found with a test for IgG antibodies. IgM antibodies tend to be more nonspecifically sticky than IgG antibodies in these assays, so one or more types may bind to proteins without true affinity for the protein. If the IgM antibody test but not the IgG test comes back positive, and other evidence for Lyme disease is not conclusive, it is important to repeat the blood test in a few weeks. In the normal course of infection, the IgM antibody level often will have fallen in the interim, and now the level of IgG antibodies will be higher. The IgM antibody test is not recommended if the infection is suspected to have been present for more than three months. Many misdiagnoses of late Lyme disease can be attributed to physicians following only IgM test results and disregarding the negative IgG test.

Antibody Testing for Neurologic Disease

When Lyme disease of the nervous system is suspected, special considerations for testing come up. This would not apply when involvement is restricted to nerves outside the brain, such as with many cases of Bell's palsy or disordered nerve function in an arm or leg. The treatment plan in such cases is usually the same as for other cases of disseminated infection. But when symptoms and signs point to invasion of the spirochetes into the brain itself, into its covering (the meninges), or within the spinal cord, the therapy usually calls for intravenous delivery of the antibiotic, which is riskier and more costly than oral antibiotic therapy (chapter 8). To justify this more aggressive course, physicians may seek laboratory evidence that the infection has extended into the central nervous system (CNS). Such evidence would be the production of antibodies locally within the CNS under the stimulus of the spirochetes present there.

The CNS is protected from direct incursions by pathogens from the outside by its enclosure in the skull and the backbone. It is also sequestered from the usual activities of the immune system. The filter from the blood into fluid bathing the brain and spinal cord—the cerebrospinal fluid (CSF)—is particularly tight. This filter, or blood-brain barrier, under normal conditions allows passage of glucose and some other nutrients the brain needs but hardly any white cells. In other organs, the white cells more easily pass from the blood into the tissues and back into the blood again. The filter to the CNS is so tight that antibodies, which are larger than most other proteins in the blood, cannot pass into the CSF either. This means that in the absence of infection or some other inflammatory disturbance in the CNS, there are few cells in the CSF, and the antibody levels are lower than in the blood.

But if there is an infection of the brain or the meninges, two conditions change. One is leakiness of that previously tight blood-brain barrier. Under the action of inflammatory cytokines and other substances, white cells can pass more easily into the CSF and brain. The second change is the beginning of antibody production within the CNS itself. The upshot is that white cells rise in number in the CSF, and there are increased concentrations of proteins, including antibodies (chapter 2). Of importance for documenting the presence of the spirochetes in the CNS is whether there is also this local production of antibodies. If there is, it means that the concentration of antibodies to B. burgdorferi will be higher in the CSF than in the blood. When the patient is healthy, it is the other way around. If it is just a question of leakiness (which may be caused by some conditions outside the brain), then the antibody concentrations in the CSF will approach that of the blood but not equal or exceed them.

The CSF for the testing is obtained by a lumbar puncture (LP), or spinal tap. This procedure is routine in hospitals and clinics. It is similar to what is done for the epidural anesthesia that women may have during childbirth or that both men and women have during some types of surgery. But without doubt, the LP is more complicated and comes with higher risk of adverse events than a simple blood draw from an arm vein. It is performed by a physician with special training. The patient lies on his or her side, curled up so that the back is flexed, which opens up the spaces between the individual backbones, the vertebrae. This allows the physician to direct a long thin needle through the space in the lower back—the lumbar region—past the protective covering to gain access to the CSF bathing the spinal cord. A few milliliters (a fraction of an ounce) of fluid are removed, and these are sent to a laboratory, where cell counts are done and protein measurements are made. If there are no or only a few white cells and the protein level is in the normal range, there would be little justification to carry out more specialized tests from that point. But if cells and protein level are abnormally elevated, then other tests besides those for antibodies may be performed. One of these is the level of a small protein called CXCL13, the elevation of which is correlated with the production of antibodies in the CNS itself.[9]

The tests for antibodies to *B. burgdorferi* in the CFS are the EIA and Western blot, which are performed as they are for the blood. The blood is also tested. The difference here is that there is an estimate of the amount of antibodies in the blood and in the CSF. It is not just a positive or negative readout of the tests. The quantification of the antibodies to *B. burgdorferi* provides a way to compare the concentrations in the CSF to those in the blood. If the concentrations are comparatively higher in the CSF than in the blood, this would be strong evidence of infection of the CNS, thus justifying intravenous antibiotics.

These steps are not commonly taken in North America. The measurement of antibodies in the CSF is a procedure more frequently performed in Europe, where the form of Lyme disease involving the brain and other parts of the CNS is more common than in North America. This difference is attributable to the different species of Lyme disease *Borrelia* on the Eurasian continent (chapter 3). There is also the higher risk in Europe of a tick-transmitted virus infection of the brain. This may produce a pattern of protein and white cell changes in the CSF that is different from *Borrelia* infection.

Direct Detection: Cultivation

There are three major limitations of diagnostic tests that aim to measure antibodies to an infecting microbe, and this applies to other pathogens in addition to Lyme disease spirochetes. The first is the time it takes for the immune systems of infected people to respond with antibodies and for those antibodies to rise above a threshold for detection by the test.

The second limitation is the tendency of some antibodies to bind to the test substances because they resemble something else or because of their inherent stickiness. This is a particular challenge with IgM antibodies, which are exactly what you would like to measure as early as possible in the infection. Although the frequency of false-positive reactions can be minimized by using only the most specific proteins in the assay, it will be difficult to completely eliminate false-positives among uninfected people. There likely will remain a trade-off between sensitivity and specificity. The third limitation of antibody-based tests is the persistence of antibodies long after the active infection has resolved. A falling concentration of IgG antibodies may indicate that the pathogen has been eliminated, but it may take weeks to months for this change to be discernible.

Given these drawbacks of the antibody tests, we might ask whether there are more direct and accurate ways to make a laboratory diagnosis of *B. burgdorferi* infection. This would be most useful for two situations: one, during early infection when the rash is either absent or atypical, and two, as assessment of whether infection is still active after treatment. In answer to our question, yes, there are more direct approaches to laboratory diagnosis, but they have their own shortcomings. For most suspected cases of Lyme disease, the antibody test has much to offer and is the least expensive assay. Still, there is a place for "direct testing" in the diagnostic process for Lyme disease in some cases. By direct testing, I mean that the microorganism itself or part of the microorganism is identified in the patient. The advantage of a direct test is that, theoretically at least, there is less chance of a false-positive reaction.*

Of all possible laboratory results, the least ambiguous is the growing of *B. burgdorferi* from a sample taken from the patient. This is the evidence that further nailed down the case that a spirochete caused Lyme disease.[10,11] Isolation of *B. burgdorferi* (or of one of the related species in Europe or Asia) is the gold standard by which other tests are judged. In most patients with early Lyme disease, the spirochetes can be cultured out of a snippet of skin taken from an erythema migrans skin rash. Such

---

* Another sort of direct detection assay is examination of a still-embedded tick for presence of *B. burgdorferi*. This approach is considered in chapter 13.

a skin biopsy is routinely performed for the diagnosis of other skin conditions, and the biopsy procedure can be done in a few minutes in the physician's office. An alternative sample for cultivation is whole blood, again best taken during early infection, the first two to three weeks. The culture of blood is more likely to yield spirochetes if there is evidence of dissemination, such as fever and multiple erythema migrans rashes, at the time of the draw. When cultivation of the skin biopsy or blood are carefully done by experienced hands, the success rate is 50 percent or higher.[12] Besides providing proof of infection, the recovery of the spirochete from the patient also allows for determining what strain it is. As discussed, some strains of B. burgdorferi are more invasive than others (chapter 3).

Unfortunately, though taking blood samples and doing skin biopsies are routine medical procedures throughout the world, the cultivation of Borrelia spirochetes is not. The cultivation of spirochetes in the laboratory requires a complex, expensive medium and patience. It takes up to four weeks for the spirochetes to grow out. Few laboratories in the world are equipped to culture B. burgdorferi. Most of the laboratories with this capability are academic research laboratories or the facilities of public health agencies, and the procedure is seldom done for routine diagnostic purposes in these labs.

Other possible sources for cultivation of spirochetes are the CSF that surround the brain and spinal cord and the fluid that lubricates the joints as they move. The latter is called synovial fluid. The synovium is the tissue lining the cartilage and bones in the joints. When there is arthritis of the joint with inflammation, there is often a buildup of fluid with increased numbers of white cells. Isolation of spirochetes from the CSF has been achieved several times, mostly in Europe, where CNS involvement is more frequent. But overall, the success rate is low even when there is other evidence of spirochete invasion of the CNS, such as local production of antibodies. The success rate for cultivation of B. burgdorferi from synovial fluid in cases of Lyme arthritis is very low to zero.

With the exception of Europe's acrodermatitis chronica atrophicans, in which the recoverable spirochetes may be present in the affected skin for months to years, there is a narrow window of time for isolating the spirochetes directly into culture. The success rate is highest when there is erythema migrans rash or during the early phase of dissemination through the blood. Effectively, this means this first month of infection. Once the spirochete removes itself from the blood to the tissues, it seldom if ever circulates in the blood again. In the deeper tissues, such as the joints' synovium and the covering around the brain, the amount of inflammation is out of proportion to the few numbers of bacteria present. One commercial laboratory in the United States reports success using an unconventional procedure in cultivating *Borrelia* from blood of patients at all stages of disease.[13] But there has not been independent validation of this method, and the published results from this laboratory have been questioned.[14]

Direct Detection: The Microscope

Another option for the biopsy specimen is examination under the microscope. The snippet of skin is immersed in formaldehyde to preserve it, and then thin slices are made. They are put on glass slides and further processed with stains to better reveal the cells and tissue structure. Skin biopsied for other reasons, such as suspected cancer, is usually handled in this way. Although this routine type of pathology exam can reveal inflammation of the skin, by itself the picture is only suggestive, not definitive, for Lyme disease. There are special stains made with silver salts that can help reveal the telltale shape and size of spirochetes in the tissue, but this evidence would not be proof that the spirochete was *B. burgdorferi*. It could be another species of spirochete. For proof that it was a *Borrelia*, the pathologist would use an antibody that is specific for *Borrelia burgdorferi* and that has been tagged with a fluorescent molecule, so that when the antibody binds to spirochetes in the tissue slices on the slide, it can be revealed under the microscope and ultraviolet light. This procedure, called a "direct immunofluorescence," was done more commonly before the introduction of the PCR test and is seldom performed today.

Direct Detection: The PCR Test

What makes cultivation of the spirochetes so difficult is that they are present only in very small quantities in most of the accessible parts of the body. A Nobel Prize–worthy technique offered a solution to this problem. If there are not enough organisms in the sample to begin with, this technique increases the number of bits of the spirochetes by up to a billion-fold. The technique is the polymerase chain reaction, or PCR test.

Of course, only by actually growing the microorganism in culture can the numbers of whole cells be increased to this extent. What PCR testing does is circumvent the need for the organism itself to multiply. PCR accomplishes this—and within hours instead of days—by taking part of the DNA, the molecules of the cell's genes, and making millions of copies of it. If only a few spirochetes, or even just one, are present in the sample, this is sufficient for PCR to do its work. The multiplication, or amplification, of the DNA is achieved in a small tube by an enzyme that copies a piece of DNA over and over again. What made this possible was the development of machines that could heat and cool very quickly, many times, measured in seconds. With each heating-cooling cycle, the amount of DNA doubles. The process continues—commonly for forty cycles—until the DNA can be easily detected in the laboratory. By starting with material for the amplification reaction that is unique to the microorganism in question, the specificity of PCR testing is assured.

The advantage of the PCR test for Lyme disease is that traces of *B. burgdorferi* can be detected in samples, such as joint fluid, that are seldom positive by culture.[15] If the specimen, such as a biopsy from a skin rash or a blood sample, was on track to be positive by culture, PCR would give a positive result in much less time and nearly as accurately.

Unfortunately, there is a downside to the PCR test: the apparatus and the enzyme will obediently amplify whatever it is presented with. The test doesn't care whether the DNA comes from a patient specimen or somewhere else. PCR is very sensitive to contamination by stray DNA molecules in the laboratory. And with billions to trillions of copies of the DNA being made during the PCR process, it is not surprising that some of the DNA molecules get into the laboratory environment. They may land on bench tops, light fixtures, walls, and the hair of the technician. They may float around on dust particles. One molecule landing in the test tube is enough to invalidate the test. The risk of specimen contamination is so high that laboratories attempting to do this testing for diagnostic purposes should be properly equipped and staffed to guard against contamination and ensure quality control. The consequence of DNA contamination of the specimens is a false-positive test: someone may be inaccurately labeled as having Lyme disease as a result.

In my first book on Lyme disease almost twenty years ago, I wrote this: "At present, PCR still has to be considered an experimental procedure. As of this writing, there is no FDA approved PCR test for Lyme disease." I am disappointed that this is still the case. Although there are FDA-approved PCR assays for a number of conditions, including infections with HIV and hepatitis C viruses, there is not one for Lyme disease. In the next chapter, I discuss some of reasons for this. The PCR assay is performed in many different laboratories—academic, government, and commercial—in the United States, but not by what could be considered a nationwide or consensus standard. Any laboratory that offers the PCR test for Lyme disease diagnosis on a commercial basis may caution that the test should be used "for research purposes" only. If the physician and the patient request PCR testing, the laboratory doing the test should be willing to provide information about their PCR test's accuracy. It is also reasonable to expect that the performance of the laboratory's test has been impartially evaluated by an outside agency.

Chapter 7

Putting Laboratory Testing in Its Place

Before the EIA

The first antibody test for Lyme disease was neither EIA nor Western blot. In fall 1981, Willy Burgdorfer and I had just isolated in culture a new spirochete from *Ixodes scapularis* ticks, collected by Jorge Benach on Shelter Island, New York. One of the first things we did was to determine if the sera of Shelter Island Lyme disease patients had antibodies to this new microorganism.[1] The test we used is performed much the same way as EIA tests are, but there are important differences. In those early laboratory tests, the results were determined under a microscope with a test called "indirect immunofluorescence assay," or IFA.*

In an IFA test, dead spirochete cells are put on a glass slide; the attached cells are exposed to antibodies in the patient's serum; the unbound antibodies are washed away; and then the bound antibody is revealed through the use of a second antibody. The second antibody has a fluorescent dye attached to it instead of an enzyme. Under ultraviolet light, the dye is a brilliant apple green. The operator can tell whether there are antibodies to the spirochetes in the serum by observing whether the spirochetes on the slide light up under the microscope.

---

*The IFA is distinguished from the direct immunofluorescence assay (chapter 6), or DFA, by its use of the second antibody. Whereas the DFA is used to directly detect the spirochetes in ticks or in animal tissues, the IFA is used mainly to detect and measure antibodies to the spirochete.

One advantage of the IFA test is that the results are quantitative rather than the qualitative positive or negative of routine EIA testing. The IFA results are reported as the highest dilution (usually in a twofold series, which doubles each time, as in 1:10, 1:20, 1:40, etc.) at which fluorescent spirochetes were still visible. This result could be compared with the result from a follow-up blood sample taken a few weeks to months after the completion of antibiotics, to assess whether the antibody levels have decreased, an indication of successful therapy.

The IFA also has the advantage of allowing the microscopist to see the binding of antibodies to actual cells under the microscope. In the fluorescent patterns, there may be subtle but nonetheless telltale signs indicating that the antibodies bound are true antibodies to *Borrelia burgdorferi*. The disadvantage of the IFA is that it requires a fair degree of skill and experience to look at the slides and interpret the results accurately. It is easier to train a technician to do EIA than IFA, and many samples can be run simultaneously by EIA. That explains why the much less costly EIA tests far outnumber IFA tests.

But for the first couple of years after the discovery of *B. burgdorferi* in North America and other species in Europe, the IFA test was the dominant test for antibodies in people with suspected Lyme disease. We provided that first isolate, named strain B31 (short for the lab jargon name "Burgdorfer, Benach, and Barbour's first isolate"), to other researchers and diagnostic companies for use in tests they developed. There was no patent or other proprietary restriction on the organism, so no single company could tie up the technology. The B31 strain is now widely used as the basis of whole-cell tests. This strain was also the first to have its genome sequenced.[2] In the complete catalog of genes found were hints of other, potentially useful proteins. Some of these are in development for second- and third-generation antibody tests.

A Home-Brew Western Blot Assay

What came to be called the Western blot procedure was described first in a scientific journal in 1979. The "Western" epithet followed a convention of applying the adjectives "Southern" and "Northern" for analogous blots of DNA and RNA (that is, ribonucleic acid), respectively.* Soon after isolating the Lyme disease spirochete in culture and observing that it could be passed from one tube of medium to another without dying off, I used some of the purified cells for a Western blot with two serum samples from a Shelter Island resident who was a patient of Dr. Edgar Grunwaldt's.[3] One sample was obtained in June at the time the individual had erythema migrans, fever, muscle aches, and malaise. A second blood sample was from March of the next year when the patient was free of symptoms. I prepared a radioactively labeled compound with the property of binding to various sorts of IgG antibodies for the same objective. Exposing the blot to x-ray film for a few minutes in the dark and then developing the film yielded images where the radioactivity was highest. These corresponded to the bands nowadays revealed by the enzyme in the blot assay. With the first serum, there were no bands. With the second, or convalescent, serum, there were eleven bands, giving the appearance of a barcode to the exposed x-ray film. With that result, Willy Burgdorfer and I became more convinced that we were on to something. This finding complemented what we saw with the IFA. This patient's convalescent serum IFA dilution was 1:1280, one of the highest we had observed.

In subsequent Western blot studies at Rocky Mountain Laboratories, we and our two collaborators, Allen Steere and Edgar Grunwaldt, observed that sera from other patients with Lyme disease produced similar patterns of bands. Moreover, as the disease progressed from erythema migrans to Lyme arthritis, more and more bands appeared in the blots.[4] Although no two patients had identical blot band patterns, several bands were common to all or most blots of Lyme disease patients. We also examined the reactions of many other serum samples from individuals without a history of Lyme disease. These included people who were reportedly healthy as well as people with other diseases that had some

---

*The "Southern" tag originated from the last name of an author of the paper first describing the DNA blot technique. But there was no "Dr. Western" or geographic significance in this context. "Western" here is a play on words.

resemblance in clinical features to Lyme disease, such as arthritis from other causes.

Though most of these non–Lyme disease sera gave no hint of bands on their blots, some control sample blots had one or a handful of bands. For the most part, these were not the same as the bands that were associated with having Lyme disease, but some control bands were shared with Lyme disease patient bands. These bands would not be informative— and could be misleading—and on this account were excluded from the diagnostic set.

Over the next few years, many other investigators and labs carried out Western blot studies, and the assay became available commercially. Manufacturers of some kits obtained FDA approval for the test, but there was no standard across different laboratories and different test versions. Something had to give if there was to be a common reference for this diagnostic test. A 1993 paper by Frank Dressler, Allen Steere, and colleagues proposed a set of criteria for a positive blot result.[5] This became known as the Dressler criteria. The identities of the ten proteins in the set plus the requirement for at least five of the ten bands being present were largely what was adopted at the Dearborn, Michigan, conference the next year (chapter 6). (These criteria are more commonly called "CDC criteria" today, but this label mistakenly attributes to the CDC more authority in the realm of diagnostic test regulation than the agency would claim or in practice has.)

Achieving a consensus opinion about the criteria for a positive Western blot was progress; few would want to go back to the more freewheeling days when criteria varied by lab. Nevertheless, as in the IFA to EIA transition, something was lost as well as gained in going from a more nuanced if idiosyncratic interpretation of the blot pattern of bands to the more regimented by-the-numbers approach of the Dressler criteria. One of the enduring controversies in Lyme disease is Western blot interpretation, specifically which bands to include in the set. Most involved with this question agree that the only point in doing a Western blot or its equivalent is if there are two or more proteins that count for diagnosis. If there is just one protein, like VlsE, an EIA will suffice. But which are the best set? Including additional proteins in the set might raise the sensitivity of the blot assay (which as practiced now falls short of ideal) but perhaps at the cost of specificity. For example, one of the proteins in contention is the OspA protein, one of the surface proteins that we will hear more about in chapter 13. OspA was one of the proteins that we identified as being commonly reactive among Lyme disease patients with Lyme arthritis. But it is also the most abundant protein made by the spirochetes in culture. Not infrequently, blots of control sera will have faint bands at the OspA level. Is this something to take seriously or is it more trivially attributable to it being present in higher amounts than other proteins on the blot? Until we move to a system in which purified recombinant proteins are painted on blots and in equivalent amounts, the debate may go on. Research progress toward a second-generation blot comprising recombinant proteins is being made on both sides of the Atlantic, but it seems to be easier to gain regulatory approval for these in Europe than in the United States.

Laboratory Dependability and Quality Control

As the testing for antibodies moved from scattered research and public health laboratories to wider application in a variety of hospital and clinic laboratories, regional and national commercial laboratories began offering IFA, EIA, and Western blot tests. Then manufacturers began producing the test kits that were being used in these more centralized laboratories as well as for individual hospitals and clinics. There are still some academic and public health laboratories that produce their own test materials and carry out the assays themselves, but only a few. The majority of Lyme disease antibody tests in North America are performed in large commercial laboratories with a nationwide or regional clientele of hospitals, clinics, and individual practitioners.[6]

The testing laboratories have to meet specific standards and achieve certifications, especially if interstate commerce rules apply and if Medicare or another large health care plan is involved and would pay for all or part of the cost of the test. One common certification is called CLIA, short for Clinical Laboratory Improvement Amendments. CLIA is under the purview of the Department of Health and Human Services' Centers for Medicare and Medicaid Services, which covers 100 million people through Medicare, Medicaid, the Children's Health Insurance Program, and the Health Insurance Marketplace. The CLIA certification, or accreditation, means that the laboratory has been evaluated with regard to a series of standards and criteria and is approved for doing diagnostic testing. It does not mean that the individual tests offered by the laboratory have been evaluated or approved. A CLIA-approved laboratory may offer tests of little or no value. Another commonly sought and valued accreditation for a clinical laboratory is from the College of American Pathologists (CAP). This is an evaluation by the major professional society for pathologists, the medical specialists who operate most clinical laboratories. This accreditation involves on-site inspections.

The large national and regional laboratories, such as Quest Diagnostics, LabCorp, and ARUP Laboratories, are expected to have both CLIA and CAP accreditation. They would not likely produce the Lyme disease test kits in-house but would instead purchase them in bulk from one of various manufacturers of the test's materials and reagents, such as blot strips with the proteins already applied, and vials with the second antibodies and solutions for performing the reactions. The test kits commonly come with a positive control, that is, a serum from someone (or a pool of sera from many people) with confirmed Lyme disease, as well as a negative control. There may be a key to the blot strips that indicates where the ten diagnostic bands are located, so the technician can just line up the patient's blot with the key to determine if a band is present or not. There are some exceptions, such as the new EIA based entirely on recombinant proteins, but for the most part, the test kits for the EIA and Western blot using whole cells are a commodity, and the manufacturers compete on price. This is also the case for EIA tests based on the entire VlsE protein or the C6 peptide, since no one manufacturer has an exclusive license to sell it. Several manufacturers have received FDA approval for their versions of the EIA and Western blot test kits for marketing. These approvals usually have been awarded under what is known as a "501(k) clearance," upon the FDA's review of the submitted evidence that the test is, in its jargon, "substantially equivalent" to (that is, at least as effective as) an already approved test on the market.

One salutary consequence of this trend toward centralization of testing and commodification of the tests themselves has been a reduction in the discrepancies between laboratories in their test results for the same patient sample. Before 1995 and publication of the Dearborn recommendations, there were no national standards for how the tests were to be done. Even with standards, for some time, many more different laboratories performed the tests than do so now, often according to their own methods and with their own materials. When the same serum samples were sent to different laboratories, which performed them blindly, the variation in the results between laboratories was discouraging. This is what I wrote in 1996: "Many experienced and currently disgruntled physicians view the available Lyme tests as not being worth the trouble. In fact, in some states the numbers of Lyme disease tests ordered have decreased. After their initial enthusiasm about finally having a blood test for this condition, many physicians are now unsure how to interpret a positive or a negative test."

The overall situation has improved since then. That's not a claim, though, that variability has been eliminated. With each step in the process—from making the medium, to growing the bacteria, to preparing the cells for EIA wells or blot strips, to performing the assay procedure itself, and finally to taking measurement—mistakes and random error can occur. The test kit manufacturers and the large commercial laboratories carry out quality control assessments periodically to minimize mistakes and errors. And if one manufacturer's kits began to decline in quality for a customer, there would likely be more than one competitor to take its place as a supplier. But to get close to 100 percent reproducibility would probably price the test kit out of the market for that manufacturer. So we have to accept some variability in results if the same sample is tested twice. If this is done in the same EIA plate, say, as a duplicate sample in a different well, the variation is generally less than 5 percent. As an example, if the raw reading is 0.75 for one of the wells, it would be no more than 0.79 or less than 0.71 for the duplicate in another well. But if the same serum sample is subjected to different runs, for example, one on Monday and another on Thursday, or with different lots of test reagents, the variation may be 10 percent or so. It may be 15 percent or so between different laboratories performing the same type of test kit. There tends to be more variation between different runs of the same sample for EIA tests for IgM antibodies than for IgG antibodies because of the inherent lower specificity of the binding of IgM antibodies.

Can we live with this? Perhaps there is a niche market for a boutique laboratory that competes via claims of better quality rather than price. Some private laboratories offer testing on a cash- or credit-card-only basis. But for most patients, this not going to be an option. Both physician and patient should have eyes open to the possible consequences of variation in test performance. If there is a 10 percent variation in the reproducibility of the results for the same sample—let alone what it could be for different samples from the patient—for a serum that has antibodies to *B. burgdorferi* but not in particularly high levels, the EIA test could be interpreted as positive one day and negative on a run another day.

How might that play out in a real-life situation? Imagine that for the first run of a patient's blood sample, the raw EIA value was 0.34, and the negative control value for that run is 0.11. The laboratory's criterion for a positive test is an index of 3.0. So in this instance, the patient's index at 3.1 (0.34 divided by 0.11) is above the cutoff, and accordingly, the interpretation of the sample is positive. On another day, for the same sample in the same laboratory, the raw value comes out to 0.32, about 7 percent lower, while the control serum's value is 0.12, about 9 percent higher. Neither is much different from the first run and both are within the range of expected values under these conditions. The index is 2.7, a negative result. Same serum, two different outcomes. That's the problem with the reporting of result as either positive or negative, and why some healthy skepticism is appropriate when the laboratory test results are at odds with the clinical diagnosis.

(Another possible outcome for a serum with antibodies at that borderline concentration, where results can flip between positive and negative, is a report of an "equivocal" result. That would be reasonable grounds for going ahead with a Western blot of the serum, or obtaining another blood sample in about two weeks and running another EIA.)

Differences in Strains and Species

A frequently proposed alternative explanation for an unexpectedly negative test result is that the wrong type of *Borrelia* cells were used in the assay. By this account, the patient had Lyme disease, but the test was negative because the assay was formulated with cells that were different from what the patient was infected with. Most of the assay kits for either the EIA or Western blot in North America use only one strain of one species. A commonly used cell for the whole-cell-based assays is strain B31 of the species *B. burgdorferi*, or another strain, such as N40, that may be used in its place. But as we know (chapter 3), other species of *Borrelia* can cause Lyme disease. This is not an issue for Lyme disease acquired in North America, where to date *B. burgdorferi* alone has been the etiology. An assay based on just that species is suitable for suspected Lyme disease acquired on that continent.[7] But the species of the cells is important for antibody tests in Europe, where at least two additional species, *Borrelia afzelii* and *Borrelia garinii*, cause the disease. That's why some tests in Europe include two or more species. Which species are included in the test kit is also relevant for North Americans suspected to have acquired the infection while in Europe.[8]

What about differences in strains in the same species? How much does this affect the outcome of the EIA and Western blot tests? In chapter 3 we considered these differences and how some strains are more likely to cause more invasive infections than others. We also saw that the strains in the upper Midwest and California differed from those in the northeastern United States. For whole-cell-based assays, which strain is the basis of the test does not appear to make much difference when the outcome is either positive or negative. Most of the proteins of the different strains are so similar to one another that the antibodies would react with the same proteins of different strains. But there are more substantial differences in a few proteins between strains. One of these proteins is OspC, which may differ by more than 30 percent in amino acid sequence for a pair of strains. This is enough difference for some antibodies to one protein to fail to bind to the other protein. Assays based on a selection of more than one OspC protein or a set of other proteins that distinguish between strains are under study but not widely available.

Some Unproven Laboratory Tests

Before starting to pull together these themes on testing, we should consider some less conventional laboratory tests for *B. burgdorferi* infection. One of these is the "Lyme urine antigen assay," or LUAT. This direct test is based on the work of some researchers who claimed to find *B. burgdorferi* proteins, or antigens,* in the urine of experimental animals and some patients. This is conceivable, and similar urine antigen-detection assays have been used with modest success for some other infectious diseases. But one inherent problem with this test is that it depends on antibodies to detect the antigens. The same types of cross-reactions that plague the antibody tests can occur in this situation. In a version of the urine test that was made widely available, there was an unacceptably high number of false-positive reactions or unacceptably low sensitivity.[9] Few laboratories perform it now.

Another test that enjoyed a brief heyday of interest is the "lymphocyte proliferation test," or "lymphocyte transformation test." Just as some antibodies are uniquely directed against one type of microorganism, some specialized lymphocytes are selective in what they recognize. The numbers of this particular kind of lymphocyte, called a T-cell, increase substantially upon exposure to the microorganism in question. Accordingly, if there is an immune response to infection, then it is reasonable to think that there may also be lymphocytes that are specifically directed against something in or on these spirochetes. In the lymphocyte proliferation test, white blood cells from the patient are exposed to parts of *B. burgdorferi*, and the response of the cells is measured in terms of their metabolic activity. If the cells become very active, that is an indication that lymphocytes in the person's body recognize the spirochete.

---

*An antigen is a substance that is the target of an immune response. Although most antigens are proteins, they can also be certain sugars or some other nonprotein compounds.

But this type of test has never been routine, since it is more complex and costly to perform than antibody tests. Few laboratories offer the test, and those that do usually offer it only on a research basis. False-positive reactions can also occur with lymphocyte tests, as they do with the antibody tests. As a panel of experts concluded, the lymphocyte proliferation test in its past and current incarnations does not appear to offer any advantage over tests for antibodies.[10]

Thinking Critically about Testing

In this section, we return to how the cutoff point for positive and negative tests is determined. Simply put, the strategy is to maximize the number of true *B. burgdorferi* infections the test picks up and to minimize the number of false-positive reactions. A test's effectiveness in detecting true infections is called its "sensitivity," and its ability to distinguish present or past true infections from absence of infection is called its "specificity." On the one hand, if the cutoff point is set too low, everyone with *B. burgdorferi* might be identified but many noninfected people might also have positive sera. The specificity of the test would be low. On the other hand, if the cutoff point is set too high, many infections will be missed. The compromise is a cutoff point that might detect all but five out of one hundred *B. burgdorferi* infections while erroneously calling positive three sera out of one hundred from people who have never had the infection. In this case, the sensitivity is 95 percent. A specificity of 97 percent is obtained by subtracting the false-positive rate, here 3 percent, from 100 percent. (The test could be either the "two-tier" combination of first the EIA and then the Western blot for positive or equivocal results, or a stand-alone test, such as an improved EIA with purified proteins, with a performance equal to that of the two-tier combination.)

The odds that the test will correctly identify an infection are very good, even excellent—but only under one circumstance: when the chance, called the "pretest probability," of having the disease is high to begin with. When the history and the physical examination suggest that the chance of Lyme disease in a given person is low, then the test's accuracy—what is called its "predictive value"—is lower. That sounds obvious and even undeserving of mention, but the point is important.

The trick is in deciding what to take seriously. Should it be equal odds, 50 percent? A 10 percent chance? A 1 percent chance? In one circumstance, the patient lives in an area known to have Lyme disease and has had in the past what sounds like a rash compatible with erythema migrans; in addition, the presenting symptoms are compatible with late Lyme disease. Before ordering an EIA test, the physician estimates that the odds of the patient really having Lyme disease are three out of four. Another way of putting it is that the "pretest probability" of Lyme disease is 75 percent. If there are a hundred such patients with this pretest probability, then seventy-five would actually have the disease. For this situation and the next, assume that the EIA test the physician orders has a sensitivity of 95 percent and a specificity of 95 percent. These are probably realistic estimates for the testing as it is done in practice and not under the more careful conditions of a research study. If one hundred people with these prior odds had this EIA test, there would be seventy-one true-positive tests (75 multiplied by 0.95) and only one false-positive test (25 multiplied by 0.05). The true-positive results far outnumber the false-positive ones. In this situation, the laboratory test has confirmed the physician's diagnosis. Taking an antibiotic is a reasonable choice for this patient. There is little chance that this treatment would be carried out in vain.

In the second circumstance, the pretest probability of Lyme disease is lower. Perhaps the patient has symptoms suggestive of Lyme disease but lives in an area with few or no documented cases of *B. burgdorferi* infection. Or the patient resides in a region with documented Lyme disease but has only nonspecific symptoms such as fatigue and some joint aches. The physician thinks that the odds of the patient actually having Lyme disease are one out of twenty, that is, a pretest probability of 5 percent. Under this circumstance, the report of a positive ELISA test from the lab is less helpful, because it is equally likely that the result is a false-positive as a true-positive reaction. The post-test probability is only even odds. Half the time, the patient would not be expected to benefit from Lyme disease treatment. If we assume that the prior odds, or pretest probability, of true *B. burgdorferi* infection are even lower, say, one out of one hundred, then most of the time a positive blood test would be inaccurate.

The risk of a false-positive test is even higher if more than one test is performed at the same time. For instance, the same serum is tested not only for antibodies to *B. burgdorferi* but also to other pathogens that the ticks may transmit (chapter 11). With each additional test added to the menu, the chances of a false-positive for at least one of them go up correspondingly.

The Phenomenon of Seronegative Lyme Disease

The other side of the diagnostic coin from false-positive is false-negative. That is, the laboratory test result was negative though the patient actually had an active infection. If the physician thinks that late Lyme disease is highly likely, and if the EIA test has a sensitivity of 96 percent for late Lyme disease, the chance that someone who actually has *B. burgdorferi* will have a negative test could be 4 percent, or one in twenty-five.

A prevailing concept about infection and immunity in general is this: when microorganisms are invasive enough to cause sickness of a human being or other animal, detectable antibodies are almost always produced in response to that infection. The obvious exception would be the case of someone who has a defective immune response. Does this concept also hold true for Lyme disease? The large majority of Lyme disease experts at academic and public health institutions in North America, Europe, and Asia would agree that it does. But there is not universal agreement on this point. Some physicians and other health care providers, as well as some laypeople, believe that "seronegative Lyme disease"—meaning late *B. burgdorferi* infection with negative antibody tests—is more common than the preceding discussion, and most Lyme disease experts, would suggest.

The fact is that a sensitivity of less than 100 percent for a Lyme disease antibody test leaves the door open for speculation about a higher prevalence of seronegative Lyme disease than public health statistics acknowledge. It could also raise doubts about the prevailing paradigm of infections in the case of Lyme disease. But let's unpack this set of false-negative results for cases in which the infection had gone on long enough for an antibody response to occur, if it was going to occur at all.

We considered the trade-off between sensitivity and specificity: the highest achievable sensitivity may be at the cost of lower specificity. This implies that some patients with an inherently lower antibody production to the *B. burgdorferi* organisms (and not necessarily to other infections) will have levels that fall short of the cutoff for a positive reading. Humans who are otherwise healthy vary in the strengths of their immune responses and in what they respond to. It may be unrealistic to think that a simple inexpensive antibody test can capture the full range of human diversity while retaining a high level of specificity. Higher sensitivity could be achieved by using more than one dilution of the sera for the EIA or by going back to the IFA, but this would at much higher financial cost for performing the test.

Other possible explanations for a negative antibody test for Lyme disease are these: (1) the clinical diagnosis of Lyme disease in a particular "seronegative" case was wrong, and there was no active infection of any sort; (2) the patient was infected but with another microbe that caused a disease resembling Lyme disease; or (3) the laboratory made an error in its report, either through a mistake in performance or inherent lack of precision of the assay.

All of these alternative accounts for the false-negative results either have been documented or are reasonably plausible. We likely will have to live with antibody tests that are less than perfect, that have shortcomings, for some time. This does not mean throwing the proverbial baby out with the bathwater. There is a valued place for laboratory testing. To return to the days when only the history and an unenhanced physical were at the physician's disposal is unthinkable. But we need not equate the official case definition for Lyme disease, which for good reason is stringent about keeping track of its temporal trends and distribution for public health purposes (chapter 4), with an actionable diagnosis and treatment plan coming out of an encounter between an individual physician and an individual patient. Most of the time, the official case definition and a more customized diagnosis will be congruent, but sometimes they will not be.

DIY Lab Testing

A person without medical training or license can obtain the same Lyme disease tests that their own physician might prescribe. An individual layperson initiates the process and names the tests to be done. One example of a company that offers this service is Private MD Labs, which for about $60, as of mid-2014, will perform a serum test for "Lyme disease antibodies, total and IgM responses." According to the company's website, the request is reviewed by a licensed physician who authorizes the test. The individual selects a site from a list of nearby facilities to have the blood drawn, goes there with the prescription for the test downloaded from the Internet, and then the blood sample is sent to one of the national or regional reference laboratories where the test is performed. The customer receives the report of the results with a short interpretation of what they mean.

The decision to have the testing is no longer solely at the doctor's discretion. It is in your hands if you want to forgo or put off professional advice. This may be empowering but is not without its downsides. As discussed, many of the symptoms of late Lyme disease can be confused with other conditions, so the history of exposure becomes very important. If all the people with symptoms but without a history of exposure were tested, the majority of positive results would be false-positive. Or a person may have a positive lab test for antibodies, but the positive test is the consequence of a past resolved infection, and there is some other explanation for the current medical problem.

Ordering the Lyme disease test on one's own may in the end yield relief if it comes back negative (unless, of course, the blood was obtained too soon into the infection for antibodies to have formed). But if the report is positive, the do-it-yourself ethic will only get you so far. Is it evidence of an active infection or a resolved relic from the past? That may be tricky to sort out. If it is thought to be active, then antibiotic therapy is indicated. Obtaining antibiotics through the web and foreign pharmacies is a less-sanctioned and riskier action than getting a test performed. For most people, receiving antibiotics (or not) means paying a visit to the doctor. Antibiotic treatment is the subject of the next two chapters.

Chapter 8
Antibiotics and Lyme Disease
The Basics

■ **Case J.** A 42-year-old woman and mother of two sought medical attention during her thirty-fourth week of pregnancy for a swollen and painful left knee. She had spent time the previous summer at the Long Island seashore. Fluid was removed from the knee joint, bringing some relief. There was no evidence that she had rheumatoid arthritis or other autoimmune disease, but both the IgG and IgM tests for antibody to *Borrelia burgdorferi* were positive. She was treated with oral amoxicillin until delivery of a healthy full-term baby. There was no evidence of abnormalities or infection of the placenta. The umbilical cord blood had IgG antibody but not IgM antibody, an indication that the infant had not been infected as a fetus.[1]

■ **Case K.** A 30-year-old resident of Iowa was admitted to the hospital because of a brain seizure and confusion. She had an intravenous plastic tube (catheter) in place from the shoulder into the heart for long-term treatment of "chronic Lyme disease," a diagnosis that was made despite several negative antibody tests and PCR tests of the blood for *B. burgdorferi*. The patient had first been treated with several months of the antibiotic doxycycline and then alternating courses of various other intravenous and oral antibiotics. The catheter was placed two years previously for the administration of intravenous antibiotics for prolonged treatment. One day after admission, the patient had a cardiac arrest and died. The autopsy revealed a large growth of a fungus related to the *Candida* yeast at the end of the catheter. This fungus was blocking blood flow into and out from the heart. There was no evidence of Lyme disease on the postmortem examination.[2]

How many people in North America and Europe have received an antibiotic for Lyme disease? There are no official figures, but my back-of-the-envelope calculation suggests more than two million people, counting back to the early 1980s. If antibiotics were not effective as therapy, we would have heard about it by now. Remember those cases of disabling arthritis that prompted the 1970s investigation around Lyme, Connecticut? They continued to occur while Lyme disease was thought to be a viral infection and was left untreated. Advanced Lyme arthritis like that is hardly seen anymore,* since antibiotic treatment became the norm. And those historical impressions are backed by evidence from controlled clinical trials of antibiotic efficacy. Both sides of the argument about Lyme disease agree that antibiotics are the cornerstone of treatment. The disagreement is about who should get antibiotics and for how long.

The case histories above introduce the topic and illustrate two points at the outset. The first is that oral antibiotics are effective for late as well as early infection, but we have to consider other factors in the choice of the antibiotic—for example, pregnancy. The second point is that antibiotic treatment is not without risk of complications or adverse effects, especially when given by vein (intravenously) and for extended periods. The fatal consequences for Case K are tragic under any circumstance but especially regrettable if the diagnosis was wrong. We return to these and other Lyme disease–specific considerations after a breakneck tour of the pharmacology of antibiotics.

How Antibiotics Work

The narrow meaning of "antibiotic" is a chemical produced by one living organism that kills or otherwise inhibits another type of organism. The discovery of penicillin followed the chance observation of a mold (a *Penicillium* species) that contaminated a petri dish and inhibited the growth of bacteria around its colony. Bacteria as well as molds and other

---

*Lyme arthritis may still occur when early disease is overlooked or when antibiotic treatment is declined, as a 2012 case in California illustrated.[3]

fungi can produce antibiotics. A broader definition of antibiotics includes chemicals that have been created from scratch in the laboratory. Sulfa is such a drug and has been in use longer than penicillin. Most of the newer antibiotics are either completely synthesized in factories or are greatly modified versions of an antibiotic originally produced by a living organism.

By either definition, antibiotics menace bacteria,* and in several ways. Some, such as penicillin, prevent the bacterium from constructing a sturdy wall around itself. In all of nature, only bacteria have cell walls of this structure, and thus these antibiotics are not likely to interfere with the growth of human or other animal cells. An attractive feature of the antibiotics that affect the cell wall is that they usually kill the bacteria outright. When the cell walls of the bacteria are weakened, the bacteria literally burst at the seams. The numbers of bacteria can drop a hundredfold, a thousandfold, or even a millionfold within the first day of treatment. A drawback of most antibiotics in this group is that they have this lethal effect only when the bacterial cells are dividing. If the bacteria are alive but are not growing in the body, antibiotics such as penicillin may fail to eradicate them expeditiously. This can be a significant factor for decisions about the duration of treatment of a slow-growing bacterium such as *B. burgdorferi*.

Other kinds of antibiotics do not kill bacteria directly. Instead, their primary action is to stop the bacteria's growth. For some bacteria, growth inhibition leads to self-destruction and death, but this path is not inevitable. One antibiotic with this type of action is tetracycline, which was the first of a larger group that includes a mainstay of Lyme disease therapy, doxycycline. Another example of an inhibitory antibiotic is erythromycin, which was the first of another class of antibiotics called macrolides. All of these antibiotics inhibit bacteria by interfering with one or more important steps in the bacterial cell's life and growth. Some antibiotics in this category prevent the bacteria from making pro-

---

*This is not by human design; antibiotics are natural weapons secreted into their immediate surroundings by some bacteria and fungi against competing microbes.

teins; like any other living creature, a bacterium without protein synthesis cannot become bigger. Other antibiotics halt the production of
DNA, the cell's blueprints; without new DNA, one bacterium cannot
become two.

While antibiotics are preventing the numbers of bacteria from getting
out of hand, the immune system of the patient is mobilizing white cells
and producing antibodies. That is, simply by stopping the growth of the
invading microorganisms, the antibiotics gain time for the body's defenses
to gear up. Moreover, whatever bacteria are already present may be weakened by the antibiotic by the time phagocytic white cells begin to flood
the infected area. In a disabled state, the bacteria are more easily ingested
by the scavenging white cells. If the person has low numbers of phagocytes or poorly functioning ones, antibiotics that only inhibit bacteria
may not be as effective as antibiotics that kill directly. This becomes an
important consideration, for instance, for people with cancer who are undergoing chemotherapy and may have few or no phagocytes. For people
with fully functional immune systems, however, there is usually no
particular advantage to using a lethal rather than a merely inhibitory
antibiotic.

What may be more important than killing ability for an antibiotic's
success is whether the antibiotic is capable of reaching the places in the
body where the bacteria are. An antibiotic may perform spectacularly in
a test tube, wiping out the numerical equivalent of the world's population within a few hours, but if it cannot first arrive at and then move into
the infected organ or tissue, then that antibiotic will fail.

Furthermore, every antibiotic has an effect only above a certain concentration. The antibiotic's concentration may be high enough in the
blood to kill or inhibit the infecting bacteria, but in a tissue, its level may
fall short. An oral antibiotic, that is, one taken by mouth, has to get from
the stomach or intestine into the blood and then find its way out of the
blood into the infected tissue. An antibiotic given by vein at least does
not have to contend with the rough-and-tumble intestinal tract, filled
as it is with food, a harsh acid, and bile—not to mention other medications that may compete with the antibiotic for absorption into the blood.

Once in the blood, the antibiotic does not perpetually circulate; the kidneys excrete it, and the liver may metabolize it to an inconsequential form. An example of this point—and one relevant to Lyme disease—is infection of the brain. As long as *B. burgdorferi* has not yet reached the brain, then an antibiotic's ability to reach into the brain is of little practical concern. However, if it is suspected that the spirochetes have become residents of the brain or spinal cord or inside their protective layer, then this becomes an issue in treatment decisions. Protecting access to the brain is a barrier that is restrictive in what it lets through from the blood into the brain and the cerebrospinal fluid (CSF). Some antibiotics are much better than others in penetrating the brain and the, CSF around it. Even then, the antibiotic's concentration in the brain is lower than in the blood and the rest of the body. This has implications for the dosage of the antibiotic, as discussed below.

In addition to whether it kills or only inhibits a specific bacterium, and whether it can penetrate into the affected tissue, a third determinant of an antibiotic's success or failure is whether it gets inside a person's cells. Some bacteria live only outside cells—in the blood or urine, or in the spaces between cells in tissues—and in this case, it is not important whether the antibiotic can pass into the cells. Some bacteria, however, such as those that cause tuberculosis, can prosper just as well within cells as outside of them. And some bacteria, such as those that cause human granulocytic anaplasmosis (chapter 11), only proliferate when they are located within cells. In these situations, an antibiotic that cannot reach the cell's interior will fail.

To achieve cell entry, the antibiotic must cross the membrane that surrounds the cell and then move to the location in the cell where the bacteria are proliferating. The membrane is made up of fats, and so antibiotics, like tetracycline and erythromycin, that are more soluble in fat have a better chance of crossing the membrane than do antibiotics that are more soluble in water. Other antibiotics can pass through the small holes that dot the cell membranes and let in nutrients for the bacterium. Generally, antibiotics in the penicillin group are not as effective at getting inside cells as are antibiotics in the tetracycline or macrolide groups.

The pharmaceutical market features many antibiotics that fit one of these profiles. Some are "me too" drugs; they are practically the same as an antibiotic that preceded them in gaining regulatory approval but have slightly different chemical structures, thereby justifying their unique names. But others stand out with distinguishing features that are pertinent to treatment decisions. An especially important consideration is the longevity of the drug in the bloodstream. In other words, after a dose is administered, how long will the antibiotic be at a level where it can damage the bacteria? Another consideration is the stability and uptake of the drug in the stomach and intestine when administered by mouth. If an antibiotic in the gastrointestinal tract is broken down, poorly absorbed, or affected by the presence of food, the blood concentration correspondingly diminishes. The original form of penicillin remains a very potent antibiotic for some bacterial pathogens, including the causes of strep throat, syphilis, and Lyme disease. *Borrelia burgdorferi* is susceptible to very low concentrations of penicillin—about a microgram in an ounce of water. It can be given orally, by injection (like a flu shot), or by intravenous infusion.

But in its original formulation, penicillin has two drawbacks that make it less desirable than some other antibiotics. One is the high number of doses per day required to maintain adequate blood levels: three or four during a day for oral use and at least four doses per day by infusion. The dosage requirements make penicillin a challenge for patients to take on schedule. For those needing intravenous therapy, penicillin treatment means extra hospital days instead of home therapy with an antibiotic, such as ceftriaxone, that can be administered once or twice a day by a visiting nurse. The second shortcoming of penicillin is its tendency to break down in the stomach under the acidic conditions and to be affected by the presence of food. Other antibiotics that are as effective, like amoxicillin and cefuroxime, are more stable and can be taken with food. But the greater convenience of ceftriaxone, cefuroxime, and amoxicillin comes with a cost: these antibiotics act against a greater variety of bacteria than simple penicillin does. As we will see, this has consequences for the large masses of friendly bacteria—our "microbiomes"—in our intestinal and respiratory tracts.

The Changing Landscape for Antibiotic Choice

There are several ways to match an antibiotic with a given infection. It might be done in trial-and-error fashion, by treating sick patients with various antibiotics and observing the outcomes. An easier and safer—but not infallible—way to determine whether a specific antibiotic is effective in treating a specific infection is to test the drug against various bacteria in the laboratory. This is first done by putting various amounts of the antibiotic of interest in culture tubes or in petri dishes and then seeing whether the bacteria are inhibited in their growth over the next few days.

Antibiotics are restricted to a greater or lesser extent in the types of bacteria they combat. One antibiotic may be very effective against, say, *Streptococcus*, a cause of serious skin infections and sore throats, but not against the most common agents of bladder infections, like *Escherichia coli* (*E. coli*). Another antibiotic may have the opposite activity: good for most urinary tract infections but not against strep infections. Some antibiotics were effective against many types of bacteria when they were first introduced into medical care but now are very limited in what they can do. Some are useless now for all practical purposes. This is the consequence of the ongoing evolution of bacteria into new forms that are resistant to one or more antibiotics—that is, new strains of the bacteria are not affected by the antibiotic in the way that their ancestors were. Resistance to an antibiotic usually occurs in one of two basic ways. The first is through a change in the enzyme or protein target for the antibiotic. If the antibiotic were a bolt and the bacterial protein the nut, one could say that in a resistant cell, the nut has been changed in its diameter or thread count, and the bolt no longer fits. The second way in which bacteria become resistant is by mounting a chemical counterattack that destroys the antibiotic.

Under the selective pressure of antibiotics in the hospital and community environment—not to mention on farms and ranches where antibiotics are freely used in animals—the resistant "mutants" proliferate and spread from person to person, animal to animal, and animal to person. What happened to penicillin is a good example of how the development of resistant bacteria limits an antibiotic's useful life. This one-time miracle drug in the 1940s and 1950s used to be a life-saving treatment against *Staphylococcus*, which causes boils in the skin, purulent wound, and serious blood infections. But within a few years after the introduction of penicillin, some strains of *Staphylococcus* bacteria that grew with impunity in the presence of formerly lethal amounts of penicillin began to appear and spread throughout the world. These and even more-resistant strains, such as MRSA (methicillin-resistant *Staphylococcus aureus*), now predominate.

Antibiotics That Work against Spirochetes

Physicians in Europe had since the 1950s successfully treated erythema migrans with penicillin or tetracycline, because of unsubstantiated suspicions that a bacterium caused it. Once the cause of Lyme disease was identified as a spirochete in the early 1980s, laboratory studies confirmed *B. burgdorferi*'s susceptibility to penicillins and tetracyclines. These antibiotics had known effects against other spirochetes, like the agent of syphilis and another group of *Borrelia* species that causes relapsing fever (chapter 11). Even low concentrations of penicillin and tetracycline inhibited the growth of *B. burgdorferi* in the culture media.

The Lyme disease spirochetes were also killed or inhibited in the laboratory by macrolides and some (but not all) cephalosporins. Cephalosporins, such as ceftriaxone and cefuroxime, which are commonly prescribed antibiotics for Lyme disease, are similar enough to penicillins to be chemical cousins. They also interfere with the formation of the rugged walls of bacteria. Like the penicillins, cephalosporins work best outside human cells rather than inside them. Common macrolides in use now are erythromycin, azithromycin, and clarithromycin. Macrolides, like the tetracyclines, can be effective against bacteria inside cells as well as outside. Another effective drug in the laboratory was chloramphenicol, an older antibiotic of a different class and structure. It is particularly good at getting into the brain and inside cells, but in rare circumstances, it has a deadly side effect and is seldom used in North America or Europe any longer. More recently, tigecycline, a newer member of the tetracycline group that is intended to be used for hospitalized patients, showed impressive activity against B. burgdorferi in the laboratory.

Several other antibiotics fell short in the inhibiting and killing tests performed in the laboratory. Some of the earlier versions of cephalosporins, notably cephalexin (also known by the brand name Keflex), had disappointing outcomes. Less effective antibiotics also included three other antibiotic groups: (1) sulfa drugs and the combination of sulfa and trimethoprim, which are frequently used to treat urinary tract infections; (2) aminoglycosides, like streptomycin and gentamicin, which are primarily given by injection or intravenous infusion for serious infections; (3) rifampin, a drug that is particularly good at penetrating cells and is used in tuberculosis therapy; and (4) ciprofloxacin (colloquially, cipro) and some other antibiotics of the fluoroquinolone class, which are commonly prescribed for outpatients and inpatients for a variety of bacterial infections.

Strictly speaking, B. burgdorferi is "resistant" to these four groups of antibiotics: the concentration at which there is an effect on B. burgdorferi is higher than what can be safely or feasibly achieved in humans. But this low susceptibility to the drug was inherent; it was there to begin with, before antibiotics were introduced or ever given as medicine. This is different from the resistance that is acquired as the result of mutations or uptake of new genes. The mounting worldwide concern over the diminishing options for antibiotic treatment for some important bacteria, like MRSA, is about this acquired type of resistance. The Borrelia spirochetes are by their nature little affected by ciprofloxacin at attainable concentrations in the body. The drug does not fit as well with its target enzyme in B. burgdorferi as it does, say, with E. coli's corresponding enzyme. Only when the drug is at high concentrations, at which troubling side effects can occur, does the antibiotic result in death of the spirochetes. Consequently, ciprofloxacin is a common treatment for E. coli urinary tract infections, but not for erythema migrans.

Another member of the ineffective antibiotics list is metronidazole (known by the brand name Flagyl), which works against bacteria that thrive only in oxygen's absence.* *Borrelia* spirochetes are not in that category; they need at least some oxygen to grow and divide. There are some other spirochete species that are limited to environments with no or very low amounts of oxygen, and the growth of these spirochetes is blocked by metronidazole. But both Lyme disease and relapsing fever *Borrelia* species grew in high concentrations of this antibiotic when it was tested.[4] At odds with this finding are reports of metronidazole's deleterious effects on *Borrelia* spirochetes under special stressful conditions in the laboratory.[5] This effect was also reported for the related drug, tinidazole, which is mostly used to treat parasite infections. Whether this laboratory phenomenon is pertinent for the success of this antibiotic as therapy is not known, because animal studies and controlled clinical trials of metronidazole or tinidazole for treatment of Lyme disease have not been done.

Testing Antibiotics in Laboratory Animals

Only a couple dozen published studies have evaluated the effectiveness of antibiotics in Lyme disease *Borrelia* infections of animals. Different studies used different sets of antibiotics and different experiment protocols—the doses, routes of delivery, dosing intervals, and durations of treatment.[6] Most studies used the laboratory mouse; the remaining studies were done in hamsters, gerbils, dogs, or ponies. A limitation of the studies of laboratory mice was that only one or two breeds, or strains, of mice were used. Within a given inbred strain, all the mice of the same sex are identical to one another and thus do not come close to encompassing the genetic diversity shown by human or wild mouse populations. Even the dog studies were limited to a single breed, the beagle. This means that the animal experiments may not fully represent the range of responses to antibiotics that we would see in humans. A further caution for interpreting the animal studies with respect to humans is this: in experiments

---

*When bacterial cells first evolved, there was effectively no oxygen in the atmosphere, so life without oxygen was the rule; for some of these primordial hangers-on, oxygen is actually toxic.

in which the antibiotic partially or completely failed, the researchers may not have achieved and then sustained antibiotic levels that would be expected for patients under appropriate treatment. Antibiotic levels in the blood were not measured in most studies.

In general, the penicillins, cephalosporins, and tetracyclines that inhibited the growth of the spirochetes in culture tubes succeeded in treating infections of mice and other animals. After antibiotic treatment of experimental animals, spirochetes could not be cultured from various tissues, while culturing spirochetes was still possible from untreated animals.[7] The macrolides were also effective in experimental animal infections but did not live up to the expectations created by their performances in the culture tubes. Vancomycin was another antibiotic that showed potent activity in the test tube but did not fare as well as ceftriaxone in completely ridding mice of infection.

The criterion for success for an antibiotic in most of these studies was whether viable spirochetes could be recovered by culture of blood and tissues. This remains the benchmark for judging a cure. There was also a significant decline in levels of antibodies to B. burgdorferi in antibiotic-treated animals but not in untreated controls. This is generally taken as evidence that the infection is under control; the immune system backs off as the threat recedes. But when PCR was added as a direct detection method, there were some paradoxical results; namely, the PCR assay remained positive long after samples were negative by culture. On the face of it, this finding was not surprising: PCR can measure the remnants of dead bacteria as well as live cells, and some persistence of the DNA in tissues after killing would be expected. But were the bacteria really dead or only in a suspended state, still alive in the animal but not capable of growing in the laboratory? This and related phenomena are considered in the next chapter.

Overview of Antibiotic Trials in Humans

Case J was pregnant and had Lyme arthritis. She needed therapy but with a drug that would treat the fetus as well as herself, and do so without harm to either. Why did her physician decide to treat the patient with amoxi-cillin, an oral antibiotic in the penicillin group? This decision could have been based on an informed hunch, collective experience of the doctor and colleagues, or more dubiously, on an unorthodox theory. More likely it was an evidence-based decision using data from controlled clinical trials as a guide.

In a controlled clinical trial, some patients, or subjects, are randomly assigned to receive the medication that is being tested, and other randomly allocated subjects are given either another medication, one already proven to work, or a placebo, such as a disguised sugar pill. This type of trial is also called "prospective" because it evaluates outcomes in the future, rather than recording the present state (sick or well) and looking backward—retrospectively—to sort subjects into groups. Which patients receive which pill is determined randomly and is kept a secret. The best trials are "double blinded" in the sense that neither the physicians nor the people in their care know which type of pill is given to each individual. Allowing neither patient nor physician to know who is taking what—active pill or placebo?—minimizes the risk that a conscious or uncon-scious bias will affect the outcome of the drug trial. The truth about who got what is revealed only at the end of the study. A trial may also end if it becomes obvious that there is difference—according to previ-ously agreed-on ground rules—in either the disease outcome or the fre-quency of adverse effects between the still-blinded groups. (A committee that is independent of the study leaders keeps an eye on trial data as they come in.) If along the way, either significantly more adverse effects or significantly better outcomes occur in the treatment group than in the control or placebo group, it would be unethical either to continue ad-ministrating the experimental drug to any subjects or to withhold the experimental drug from the control group, as the case may be. An early end of a trial under such circumstances indicates that the original re-search design worked as intended. It is not an ad hoc decision.

In North America, the first investigation of whether antibiotics could be used with effect to treat Lyme disease was carried out by Allen Steere and Stephen Malawista and their colleagues at Yale University over the period 1977 to 1979, before the discovery of *B. burgdorferi*.[8] The trial was neither controlled nor blinded, meaning that patients were not randomly assigned to treatments, and the physicians and patients knew what the treatment was. But it did include patients who received no antibiotic, the rationale at the time being that there was not enough evidence that antibiotics would provide more benefit than harm. Of the 113 patients with Lyme disease with erythema migrans, 58 received an antibiotic—most frequently penicillin but a few each with tetracycline or erythromycin—and 55 did not. The patients who took penicillin or tetracycline had a more rapid resolution of the rash and were less likely to develop arthritis than those who got erythromycin or went untreated.

In a second antibiotic trial, the Yale investigators enrolled as subjects patients with established arthritis of late Lyme disease between 1980 and 1982, still before the discovery of *B. burgdorferi*.[9] At the time, the arthritis of Lyme disease was considered plausibly to be the result of an immune reaction to an infection and not the infection itself. One could point to other examples of arthritis that followed certain types of infection that had occurred but then resolved. Consequently, some of the patients in this second study received a placebo in the form of a salt solution instead of the test drug in the trial: intravenous penicillin. But this trial was controlled and blinded. As it turned out, most of the people who received the intravenous penicillin had resolution of their arthritis sooner than those in the placebo group. This difference was statistically significant, which meant that the likelihood of there being no actual beneficial effect of the penicillin—that the apparent success of the antibiotic was just a nice run of luck—was less than one out of twenty, or 5 percent.

Ever since these two studies, patients in all clinical trials for active *B. burgdorferi* infection have gotten some form of antibiotic. Once the benefit of antibiotic treatment was demonstrated, it would have been unethical to withhold antibiotics.* Subsequent clinical trials have examined the relative efficacies of different antibiotics and dosing schedules. A new antibiotic or dosing schedule is tested against a therapy that has already

---

* Lyme disease, as an active infection, is distinguished from post–Lyme disease syndrome, for which antibiotics have not yet been shown to be effective (chapter 10).

proved its worth. However, given the huge numbers of patients who have received antibiotics for Lyme disease—and the many who will in the future—it's difficult not to be disappointed about the modest number of clinical trials of antibiotics for Lyme disease. There have been only a couple in the last ten years in either North America or Europe.

There are several reasons for this. Here are some in my view:

1. Most of the antibiotics used for Lyme disease treatment are generic. That is, after patent protection runs out, generic drugs are produced by two or more companies that can compete on price and less on brand name. Amoxicillin and doxycycline have been in use for decades. Even ceftriaxone and cefuroxime, big contributors to sales for their patent-holding manufacturers, under the brand names Rocephin and Ceftin, respectively, are now available as generics. There is little incentive or funding for generic drug manufacturers to sponsor additional clinical trials that would likely add little to the bottom line and may possibly reveal some unanticipated shortcomings.

2. Once a drug receives FDA or other regulatory approval for marketing and use in patients, it is often used "off-label"—in other words, for indications that were not part of the original application by the pharmaceutical company or what the FDA stipulated. Some drugs, such as cefuroxime, list *B. burgdorferi* or Lyme disease as indications in their official "prescribing information" sheets, but many, like Vibramycin, one of the original brand names for doxycycline, do not.

3. In comparison to other types of drugs, such as those used by patients their entire lives after diagnosis with a disease such as hypertension, HIV infection, or diabetes, or for serious conditions like cancer and life-threatening infections, the market for oral antibiotics for Lyme disease treatment is small. If a new antibiotic, still under patent protection, showed promise in treating Lyme disease with a single or a handful of doses, each pill fetching $100 retail, there might be motivation for sponsoring trials, but only if the drug could also be used for more common medical conditions,

like staph or strep infections of the skin, urinary tract infections, or bronchitis.

4. Research grant-funding agencies, both public and private, seem disinclined to support further clinical research studies of Lyme disease treatment on an appropriate scale for answering the outstanding questions. Perhaps the leaderships that set funding priorities or the peer-review panels that evaluate applications view the important issues as settled.

Outcomes of Controlled Trials and Other Human Studies

With a perspective that the glass is half full, not half empty, a review of what clinical trials there have been is reassuring. The cumulative experiences of hundreds of physicians in Europe and then in North America did not mislead. Antibiotics work for active Lyme disease.

Overall, doxycycline, amoxicillin, and cefuroxime by mouth over periods of 10 to 21 days—most commonly 14 days—were indistinguishably effective as treatments for erythema migrans and early Lyme disease.[10] About nineteen out of every twenty people treated with these antibiotics for acute *B. burgdorferi* infection benefited from the treatment. Their rashes faded sooner with antibiotics than they would have without. Cultures of skin biopsies at the sites of erythema migrans turned negative with antibiotic therapy.[11] People began to feel better earlier when they were treated than when they were not. There was a significantly lower risk of late arthritis or neurologic disease if early disease was treated with antibiotics. The reported results with doxycycline, amoxicillin, and cefuroxime generally were better than what was observed in the first trial in the late 1970s, which used antibiotics, like penicillin, that had four-times-a-day schedules, which are more difficult for patients to adhere to, and that were given for only 7 to 10 days.

The older macrolide erythromycin and the more conveniently dosed azithromycin and clathromycin were less effective in the clinical studies, as noted in the animal experiments. But macrolides were not so inferior as to be ruled out as alternatives in the case of contraindications, like pregnancy, young age, or allergy, to the first-line drugs. The macrolides are preferred over more risky options, like tigecycline, or insufficiently tested newer antibiotics coming on the market.

People with evidence of dissemination, such as multiple rashes, Bell's palsy, or fever, at the time oral antibiotic treatment starts fared much better than did those who went untreated, by the criteria of the time to symptom resolution and the risk of late disease, such as arthritis. Nevertheless, patients with disseminated infection were marginally more apt to require another course of antibiotics or to endure late disease than those with a more localized infection, such as a single erythema migrans rash and few if any general symptoms, like diffuse muscle aches.

Oral doxycycline or amoxicillin was as effective as intravenous ceftriaxone for treatment of most cases of arthritis of late Lyme disease. Oral doxycycline was also as successful as intravenous ceftriaxone for treatment of nervous system Lyme disease in Europe.[12] But other studies in North America suggested that an intravenous antibiotic (penicillin, ceftriaxone, or cefotaxime, another cephalosporin) may be required for best outcome for Lyme disease in which there is invasion of the brain or other parts of the central nervous system (CNS).

A Closer Look at One Clinical Trial

The specifics of a representative clinical trial puts some flesh on the dry bones of an abstract overview. A publicly accessible 1995 paper reports on a study comparing cefuroxime,* which at the time of the trial, 1990, was an unproven antibiotic for Lyme disease, with doxycycline, an antibiotic of established value.[13] Cefuroxime was demonstrably effective against *B. burgdorferi* in laboratory cultures and in experimental animals but had yet to be formally evaluated in humans. Study subjects were drawn from patients in primary care medical practices in Connecticut, New York, and New Jersey; sites included the historically relevant locales of Old Lyme, Connecticut, and Shelter Island, New York. About two hundred patients with early Lyme disease and erythema migrans were randomly assigned to receive either cefuroxime or doxycycline. The

---

*The full generic name of the oral form is cefuroxime axetil.

doxycycline was at 50 percent higher daily dose than is typical in the United States, but this is a dose that is often used in Europe. Overall, 15 percent of the patients had multiple erythema migrans lesions, which is evidence of disseminated infection. They took the antibiotics for at least 12 days and no more than 20 days. Follow-up examinations occurred just after treatment ended and periodically over the succeeding twelve months. Overall, one in five of the total subjects had "drug-related adverse effects," most commonly an allergic rash, sensitivity of the skin to sunlight (a well-known side effect of tetracyclines), nausea, and diarrhea.

By the end of the study, 90 percent of the cefuroxime-treated subjects and 95 percent of the doxycycline subjects had "satisfactory" outcomes, meaning either complete resolution of the rash and other clinical symptoms (like headache or fatigue) by the one-month followup. This was not a statistically significant difference between the two groups. The flip side of these results were "unsatisfactory" outcomes in 10 percent and 5 percent of the two groups. A few of these outcomes were a recurrence of symptoms within a month. The rest of the unsatisfactory outcomes were "failures," meaning no improvement in the rash or clinical symptoms by an exam a few days after treatment ended. A total of eleven subjects (or about 5 percent of the original group) continued for at least one month to have joint pains, fatigue, malaise, muscle aches, and/or headache.

Among subjects who were in the "satisfactory" outcome group at the one-month checkup, about 90 percent of the cefuroxime-treated and doxycycline-treated subjects at one year out were free of the Lyme disease symptoms they'd had to begin with. Overall, about 8 percent had some residual symptoms but were improved at the one-year mark. There were three "failures," meaning they had symptoms and signs of late Lyme disease, specifically two subjects with arthritis and one with a central nervous system disorder. All of these patients had received cefuroxime, but this was not a statistically significant difference from the doxycycline group. In other words, there was a greater than 5 percent chance that it could have come out differently: either all the failures could have been in the doxycycline group or could have been distributed between both groups. We would need a clinical trial with a much larger number of subjects to adequately test whether there truly is a higher risk of late Lyme disease with cefuroxime compared to doxycycline therapy. One would think this is an important issue to resolve—it could mean hundreds of patients being inadequately treated if cefuroxime truly has a higher failure rate. But this clinical research study has not been carried out (or at least published on).

Can *B. burgdorferi* Become Resistant to Antibiotics?

There is no reason why *B. burgdorferi* couldn't become resistant. Spirochetes are like other bacteria in possessing the raw material for evolution: a thousand or so genes, each of which has a small chance of error with every duplication during growth. Here's how to demonstrate this: take about a billion bacterial cells—what's in a half-ounce of culture medium at peak growth—and incubate them with an antibiotic. A few cells survive because each has a rare mutation that by a chance event was already in the cell's genome before the exposure to the antibiotic. Under antibiotic-free culture conditions, these mutant cells would not have an advantage over other bacteria; there may even be deleterious consequences of the mutation, such as a slower growth rate. But in the presence of an antibiotic, cells with a mutation conferring the capacity to withstand the antibiotic will survive and prosper as other cells die off, leaving the field clear of competition. Something like this was demonstrated by a research study with *B. burgdorferi* and selection by an antibiotic, coumermycin, that is rarely if ever used any longer.[14]

But this feat was accomplished in an otherwise benign environment for the spirochete cells; all their needs were met in the laboratory's culture tube, and there were effectively no predators. They were competing only against each other as the population density increased. Inside a human host, a newly arisen mutant spirochete would still have to contend with white cells and antibodies, which wouldn't care whether the spirochete was antibiotic resistant or not. And even if an unfortunate patient suffered a relapse of illness under treatment because of the selection and then growth of an antibiotic mutant,* that human host would be a dead end for the mutant. This is a key distinction between Lyme disease spirochetes and medically important bacteria in which antibiotic resistance is a present and future threat. Staphylococci, gonorrhea cocci, and tuberculosis bacilli, all of which feature significant antibiotic resistance now, are adapted to life with humans and succeed on Earth by a simple transmission scheme: person to person. An antibiotic-resistant strain that develops—either by a new mutation or by picking up a resistance gene from another bacterium—can spread, for example, on hands, through sex, or by a cough, to another host, then another, and so on. As long as exposure to antibiotics is maintained in an environment—be it a hospital, community clinic, or factory farm, the antibiotic-resistant strains will disperse and may come to dominate other strains.

In contrast, for an antibiotic-resistant *Borrelia* spirochete to eventually spread beyond that unlucky current human host to another human, a cumulative series of events, each highly improbable, would have to occur. The first hurdle would be for the mutant cells to pass into either a larval or nymphal deer tick that happens to bite the infected human. Then, if the fed tick managed to avoid detection during the blood meal and then drop off in an area where it could find new hosts, it would need to survive the winter and then either bite another human again—the chance of which is exceedingly small—or feed on one of its usual hosts, like the *Peromyscus* mouse, thereby maintaining a chain of infection. But there would be no further exposure of the mutant to antibiotics in these wildlife, so any advantage the mutant might have had in the original human host under treatment is lost. If progress through any one of those steps ceases, the future for that mutant closes down.

In conclusion, the basic set of antibiotics that worked against Lyme disease thirty years ago remain as effective now as then, and this will likely be the case in the future as well. The next chapter discusses how these antibiotics are used in practice.

---

* Relapse of Lyme disease that is attributable to emergence of resistance during therapy of a patient has yet to be reported and documented.

Chapter 9

Putting Antibiotics to Use

The Safety of Antibiotics

The salutary actions of antibiotics are complemented by a good overall safety record. Of all types of medications, antibacterial antibiotics are among the safest. This is true because of the aforementioned differences in the structures and metabolisms of bacterial and human cells. There is inherently less chance of toxicity with antibiotics than with medications that are supposed to work on human cells, tissues, organs, and systems. But antibiotics are not without risk, and possible deleterious effects should be taken into account in treatment decisions.

In developing a new antibiotic, the first priority is to determine whether a candidate compound kills or inhibits bacteria. Thousands of compounds with these activities have been discovered or produced, but most of these antibiotics now sit on the backroom shelves of pharmaceutical companies. These drugs may have performed exceptionally well in exams of killing bacteria in the test tube, but they were also, to their discoverer's or inventor's disappointment, toxic to animals. This toxicity may have been discovered in studies of human or other animal cells growing in the laboratory. The cells may have grown poorly or died, or they may have shown other signs of distress. Some antibiotics proceed to whole animal testing. The drug may perform as intended for a particular disease or condition, but treated animals may develop adverse effects, such as seizures, sick livers, or depressed blood counts, as a result of the drug. And often that's where the development trail ends.

But if the drug survives this gauntlet of laboratory and animal tests, the next step in the evaluation is a trial with human volunteers. More often than not, these are young men who are usually paid for their participation. The volunteers are given the antibiotic either by mouth, by injection into a muscle, or by injection directly into the blood. The purpose of these early tests in humans is not to see if the antibiotic works against infections—the animal tests are usually predictive of this. Instead, the principal aim is to pick up any signs that the drug is toxic to people. This is done by monitoring the volunteers by questionnaire for symptoms such as headache, nausea, and blurred vision; by physical exams for signs of disordered organ function, such as a skin rash or lack of coordination; and by blood and urine tests for abnormalities in the blood, liver, kidneys, glands, intestinal tract, or muscles.

If there are no abnormalities or there is a tolerable number of minor abnormalities after this stage of testing, the FDA customarily allows more extensive testing involving human volunteers. Additional people receive the drug in various strengths and for durations relevant to real-life therapeutic situations. Sometimes abnormalities become apparent only when larger groups of people are tested. These side effects may be fairly negligible, such as mild nausea or headache, or they may be more worrisome, such as a declining number of white blood cells. In the final stages of testing, the new antibiotic is tried out on patients in hospitals, clinics, or physicians' offices. These patients or their guardians are informed of the test and sign a consent form if they agree to participate. At this point the principal question is whether the drug works to cure the infection. But participating patients are also routinely monitored for side effects. Toxicity is still a concern.

Although the evaluation process for approval of a new antibiotic is strenuous, it is not flawless or infallible. The antibiotic might perform admirably on many counts at curing patients in the trials and with apparent safety. Nevertheless, unforeseen side effects might be detected only after the drug has been approved for sale and even larger numbers of people have been treated. Some of these adverse effects may be serious enough that either the antibiotic is withdrawn from the market or extreme caution in its use is urged. For example, an antibiotic called temafloxin, a type of fluoroquinolone like ciprofloxacin, was put on the market in January 1992 but was withdrawn in June that same year after several people taking it developed severe side effects and three died.

As another example, in fall 2013, the intravenous drug tigecycline, under the brand name Tygacil, was given the "black box label," the most severe warning for practitioners, by the FDA. Further study had confirmed the agency's 2010 finding that the antibiotic was associated with a greater risk of death than other antibacterial drugs. Tigecycline had shown early promise in laboratory and animal studies as described above as an antibiotic for intravenous treatment of *Borrelia burgdorferi* infection. But now the prescribing information label warns of higher "all-cause mortality" in patients treated with the antibiotic and advises that "Tygacil should be reserved for use in situations when alternative treatments are not suitable."

Adverse Effects and Complications of Antibiotic Therapy

These and other "postmarketing" failings notwithstanding, antibiotics generally have safety records that are the envy of the makers of other medicines. Nevertheless, accounts of adverse effects of antibiotics and complications of antibiotic therapy could fill the remaining pages of this book. Call to mind an announcer's rapid-tempo litany of warnings for prescription drugs in television commercials to imagine where this could lead. There are not only books and magazine articles on drug side effects, but the Internet offers many sources (both trustworthy and less so) of information about the strengths and weaknesses of each antibiotic that might be used to treat Lyme disease or the other infections that deer ticks can transmit (chapter 11). And there is always the detailed package insert, which can as much alarm as inform.

Case K's story at the beginning of the last chapter was a particularly dramatic instance of what can go wrong, but it wasn't an example of adverse effects as they are usually classified by the FDA or in pharmacy textbooks. For all we know, Case K had no adverse effects, such as liver toxicity or bone marrow suppression, directly attributable to ceftriaxone. The

fatal outcome was mainly a complication of receiving an intravenous antibiotic through an implanted catheter over a prolonged period. At some
time a fungus invaded, probably at the point of access to the catheter in
the skin, and got established on the catheter's plastic-like material. The
fungus, not having a cell wall like bacteria, was oblivious to ceftriaxone's
action. Bacteria instead of a fungus could have occupied that place
on the catheter—that's a common-enough complication of intravenous
therapy—but ceftriaxone has such broad activity that many of the bacterial contenders for occupying the catheter in place of the fungus would
have been killed off.

Let's look at ceftriaxone in more detail, mainly on its own account,
but also as an example of antibiotics in general. Why pick ceftriaxone?
It is the most commonly prescribed antibiotic for intravenous treatment
of Lyme disease, and for good reasons. Ceftriaxone has a long track record of success as a therapeutic agent for *Borrelia* infection of the brain
and the meninges that covers it. Just as important, it need be administered just once a day, making therapy at home feasible. But even a home-
based course of ceftriaxone therapy is much more expensive—in the thousands of dollars—than a course of oral therapy. This cost is accounted
for less by the medication itself than by all the paraphernalia, like IV tubing, sterile fluids, gloves, and automated dispensers, as well as expenses
for personnel, like the pharmacist, office staff, nurse, and prescribing physician, who work as a team to administer the antibiotic.

The drug itself may have adverse effects, the ones that the package insert warns about and the FDA keeps tabs on. Some may be associated in
particular with ceftriaxone. One is gallbladder disease, which became apparent only after the antibiotic had received FDA approval and was on
the market. At higher doses of the drug, the normally liquid bile turns
to sludge, and blockage in the gallbladder may ensue. Although the gallbladder problems usually resolved after people stopped taking the antibiotic, in some cases abdominal surgery for gallbladder removal was
necessary.

Other adverse effects of ceftriaxone are shared with other antibiotics
and other drugs. One that causes consternation for physicians is a fall in
the white cell count in the blood. Normally this is good thing, a sign of

improvement when there has been an infection, and the white cell count was elevated in response to that infection. But if the number of white cells declines below the range for normal, this fall could be attributed to different factors, including the drugs the patient is on. A white cell count dropping to dangerous levels, at which the ability to stave off and fight infections becomes compromised, may mean stopping the antibiotic or starting another in its place.

A common side effect of ceftriaxone and other antibiotics, including those taken orally, is diarrhea or loose stools. This might be from a direct effect of the drug on the functional activity of the intestines, but more commonly it is the result of an ecologic change in the intestine and its "microbiome," which comprises trillions of bacteria, representing hundreds of species. If the skin is analogous to a desert in the richness of its bacterial "flora," the several-foot-long intestine is like a rain forest. There are complex interactions between the bacteria, many types of which depend on others for their growth. Because of these complicated relationships, often involving three or more different types of bacteria, it is difficult to predict the effect of making even one alteration in the intestinal environment. The consequences may be minimal, or they may be catastrophic.

Understandably, antibiotics often disrupt this inner environment. They do not discriminate between a pathogenic microbe and a benign or beneficial microbe. The longer the list of the microbiome's species affected—the broader the "spectrum" of activity for an antibiotic—the greater the effect on that environment. Even if only one type of intestinal bacterium is affected by the antibiotic, the change in its numbers may be enough to have many consequences for other bacteria. For instance, the elimination of one bacterial type may permit a second bacterium, whose numbers had been kept in check by the first, to thrive. In larger numbers, the second bacterium may make enough toxins or other chemicals to cause diarrhea. Infrequently but dramatically, the surging bacteria cause severe diarrhea, cramps, and ulcers of the intestine. This condition can be life threatening, especially for infants and patients who are already infirm. One of the opportunistic bacteria causing severe diarrhea in some people taking antibiotics is called *Clostridium difficile*. This bacterium is a rising threat in hospitals, because the spores of the bacteria are very hardy and can be transmitted from person to person on hands and on clothing and other materials. There is a laboratory test for C. diff, as it is known for short, and the toxin that it produces. Once this particular culprit has been identified, other antibiotics, usually either vancomycin or metronidazole taken by mouth, can be used to reduce the numbers of *C. difficile* in the intestine.

Another possible consequence of normal flora die-off and overgrowth of others is a yeast infection of the vagina, another location with a diverse set of species. *Candida* yeasts are a type of fungus and are not affected by most antibiotics. When bacterial species in the vagina are killed off, the *Candida* organisms increase in numbers in the absence of competition and cause vaginal discharge and irritation. A yeast infection can also develop in the mouth—where it is called thrush—when antibiotics alter the bacterial flora on the cheeks, the gums, and around the teeth.

To close this selective survey, we consider the risks of delivering ceftriaxone, or other antibiotic, by vein over days to weeks. If intravenous therapy for more than a few days is required, a physician may implant a special kind of catheter that is positioned on the chest, shoulder, or neck and that goes directly into one of the larger veins. The catheter is more stable, less likely to come loose or get blocked, in this position than in an arm, wrist, or hand vein, and there is easier access for delivery of the medication in its IV. An IV bottle need not be hooked up continuously. This ease of access provides a more comfortable experience for the patient and allows for greater mobility. Nevertheless, no matter what amount of care and precautions are taken to restrict entry of microbes into the catheter, sometimes the defense is broached, and an organism like MRSA or the fungus of Case K gains a foothold. Neither MRSA nor the fungus would be susceptible to ceftriaxone. They can grow on the catheter material and form an adherent film that is impervious to other antibiotics that might be administered. The invading microbes from their base camp on the catheter can send showers of cells into the blood. Once a catheter becomes infected in this way, the usual remedy is to completely remove the catheter apparatus and tubing. If these catheter infections are not recognized in time, the outcome may be death or disabling complications.

Drug Interactions

As was true for adverse effects, the list of possible interactions of each type of antibiotic with other medications would fill pages and pages of text. There are so many of these potential drug interactions that even experienced physicians, nurses, and pharmacists cannot keep all of them in mind. Computer programs have been created that will identify potential interactions, and several reference books provide data on all known drug interactions. Your physician and your pharmacist have access to these materials and will almost certainly consult one of them if there are any questions about a potential interaction. You can expect, too, that they will ask you what medicines you're taking—the physician, so that he or she can make safe decisions about prescribing medications for you, and the pharmacist, so that he or she can caution you about drug interactions, or even call your physician to discuss the medications, if there's a potential for a problematic interaction. If you are not asked, you should be sure to report to your physician and pharmacist all medications you are presently taking. It's not a bad idea to carry a list of your current medications with you, especially if you see different physicians for different medical problems.

Allergies to Antibiotics

Some complications of Lyme disease therapy cannot be blamed solely on the antibiotics. In these cases, the person's own immune system, not the drug itself, causes most of the damage. One such immune response is an allergic reaction to the antibiotic. It may also be referred to as "hypersensitivity." Whether the person has Lyme disease or not is irrelevant, since an allergic reaction to this specific drug could just as easily happen during treatment of a strep throat or gonorrhea. Such a reaction does not usually occur the first time someone is treated with an antibiotic; it may happen the second, third, fourth, or umpteenth time the person takes that antibiotic or one that is chemically similar to it.

The most dangerous kind of drug allergy occurs suddenly, within minutes of receiving the drug through a vein or in a muscle or, less likely, after taking the drug orally. As with a severe bee-sting allergy, the patient's blood pressure drops and breathing becomes very difficult. If adrenaline is not administered soon after the start of this reaction, the patient may die. Fortunately, these overwhelming, potentially fatal reactions are rare; for example, severe reactions to penicillin occur about once in ten thousand injections of the drug. A related but less severe reaction is hives or welts over the body. These typically appear within a few hours of being given the antibiotic. They are usually smooth in texture, raised above the skin, slightly redder than the surrounding area, and often itchy.

Another type of allergic reaction is delayed in its appearance, occurring several days after the start of antibiotic treatment. This reaction looks similar to the rash of measles. There are small, red, slightly raised spots, sometimes over the whole body and sometimes limited to certain areas, such as the legs. Between 1 and 5 percent of people taking an antibiotic develop a rash. A person who has such a reaction usually but not always has to stop taking the drug and, if treatment has not been completed, replace it with another.

Before an antibiotic is prescribed, the physician or nurse will ask the patient questions about any allergic reactions to medicines in the past. A patient's report of a possible allergic reaction to an antibiotic steers consideration away from that antibiotic and toward another. One problem in making this determination is that many people remember having had some sort of unpleasant symptoms after taking an antibiotic. Up to 10 percent of people report being allergic to penicillin. But was this an allergic reaction or a side effect of the medication? If what is reported is nausea, diarrhea, minor dizziness, or headache, it's not likely that this was a true allergic reaction, and there is a low risk of sudden collapse from receiving the same antibiotic. An indication of serious trouble ahead is more likely when the patient tells a story of difficult breathing or a rash, particularly hives, occurring within a few minutes or hours of taking the pill or shot. Most reports of a measles-like, spotty red rash within days of starting an antibiotic are also taken as evidence of a drug allergy, even if a less serious one.

The Jarisch-Herxheimer Reaction

After starting their antibiotics, a minority of patients with early Lyme disease become sicker instead of better. This is more likely if there is evidence of dissemination, like fever and multiple erythema migrans rashes. Within one or two days of beginning treatment, the rash of erythema migrans may become redder and slightly painful or itchy. The person's temperature usually rises one or two degrees Fahrenheit, and the person may have general achiness. But this apparent setback is neither an antibiotic side effect nor an allergic reaction.

Unlike an allergy or side effect, what has been called the Jarisch-Herxheimer reaction (after the two physicians who described it) occurs only when the person has been infected with a spirochete. The person could take the same antibiotic for another type of infection without suffering the same outcome. What happens in the Jarisch-Herxheimer reaction is that the antibiotic kills a large number of *B. burgdorferi* cells, and the person responds to these dead and dying spirochetes by releasing cytokines and other compounds. This local and systemic outpouring of body chemicals leads to increased inflammation at the erythema migrans site, elevates the temperature, and creates overall achiness. Thus, the Jarisch-Herxheimer reaction, while discomforting to the patient and to the physician first encountering it, is actually a sign that the antibiotics are working and that the diagnosis of Lyme disease is probably correct. Within a couple of days, the symptoms of the reaction abate and general improvement follows.

Some patients and their physicians have reported similar but not identical reactions during the treatment of illnesses thought to be late Lyme disease or that have been diagnosed as "chronic Lyme disease." In these situations, there is usually no fever or skin rash at the time treatment begins. The reaction these patients describe is an increase in joint and muscle aches, a greater fatigue, and perhaps chilliness (which is not synonymous with fever). A heightening of symptoms more than two days after the start of antibiotics is unlikely to be such a reaction. Moreover, these experiences are usually entirely subjective (a secondary quality, as described in chapter 5). In the absence of objective (primary quality) findings such as elevated body temperature or worsening of the rash, an experience of increased aches and fatigue after starting antibiotics may doubtless be important to note but cannot be taken as confirmation of the diagnosis of late infection.

Antibiotics during Pregnancy, Breast-Feeding, and Childhood

Antibiotics are generally safe, but under some conditions, certain antibiotics should not be used. Pregnancy, as in Case J, is one such condition. Care must be taken in the use of any medication, including antibiotics, during pregnancy. Among antibiotics, the tetracycline class of drugs, which includes doxycycline, should not be prescribed for pregnant women. This class of antibiotics will stain the developing teeth of the fetus a brown or yellow color. The bones are also affected, though these effects would not be visible. Because of this effect, tetracyclines should also not be taken by women who are nursing or by children younger than nine, the age at which the permanent teeth have finished coming in.

Many newer antibiotics are also avoided during pregnancy, not so much because they are associated with known side effects in pregnant women or their babies but because the effects are not known at all. In trials for toxicity and effectiveness, antibiotics are not tested on pregnant women, primarily because of safety concerns and fear of lawsuits. Although these newer antibiotics are probably safe for a mother and developing child, there is no way to be sure. This is a catch-22 situation: the antibiotic cannot be tested on pregnant women because it has not been prescribed to them before, and it cannot be prescribed to pregnant women because it has not been tested on them before. There is a similar hesitancy about studies and consequent lack of data for some antibiotics during breast-feeding and for young children.

Picking the Right Dose and Interval

The dosage of all antibiotics should be adjusted for the weight of the patient, but it is mainly pediatricians, whose patients come in a variety of sizes, who pay attention to this detail. Adults commonly—but nevertheless ill-advisedly—get the same dose regardless of their size; whether they are 4 feet 10 inches and weigh 90 pounds, or 6 feet 8 and weigh 300 pounds, they take the same amount of oral medication. This is not the best way to treat patients, since some may be getting too much medication and others not enough.

The dosage of antibiotics should also be adjusted—that is, reduced—if either the kidneys or the liver is not functioning well, and the antibiotic concentration in the blood would be affected by this. For people with kidney disease, including those on dialysis, the dosage of most antibiotics needs to be adjusted downward according to the amount of remaining function. This is because most antibiotics are excreted from the body by the kidneys, and if this organ system is not working properly, the levels of antibiotics in the blood begin to back up and may exceed the toxicity threshold. Other antibiotics are handled predominantly by the liver; these must be given in reduced dosages if that organ has impaired function.

Once an antibiotic in the blood, tissue, or cerebral spinal fluid reaches a level that kills or inhibits the spirochetes that cause Lyme disease, even much higher concentrations of the antibiotics do not speed up the spirochetes' death. Therefore, there is no advantage to giving larger doses than is necessary. But the timing of the doses is important in making therapy as effective as possible. The antibiotic may reach the critical level, but if it does not remain above that level for a certain amount of time during the day, the antibiotic may fail. This is because, although the bacteria may be temporarily slowed in growing, and some of them may even be killed, they will regain ground or even gain additional ground once the antibiotic level falls below the threshold for inhibition.

Is Longer Better than Shorter?

Few practitioners argue about the size of antibiotic doses given to treat Lyme disease in its different stages. More controversial is the appropriate duration of therapy, especially for what is thought to be late Lyme disease, or chronic Lyme disease by other definitions. There are natural limits on treatment decisions about doses: below a certain level, an antibiotic will not work, and above a certain concentration, side effects increase. Too little antibiotic, the person remains infected; too much antibiotic, the person may get sick from toxicity. The possibilities for treatment duration, however, are literally endless, ranging from one dose to lifelong consumption. A single dose of an antibiotic cures some bacterial infections, such as gonorrhea or an uncomplicated urinary tract infection. It is effective for some forms of relapsing fever, the other disease caused by *Borrelia* species. But other diseases, such as tuberculosis and some fungal infections, may require months to years of treatment for a cure. Some people with leprosy or deep-seated, inoperable infections take antibiotics for the rest of their lives; their infection can be controlled but never eradicated. And for the great majority of affected people, HIV infection calls for lifelong treatment with antiviral medications to keep HIV in check.

Is Lyme disease closer to the gonorrhea model or to the leprosy model in terms of the length of therapy required for cure? Of these two infections, gonorrhea is closer, but strep throat may be a better analogy to early Lyme disease. *Streptococcus* infection of the throat and tonsils is effectively treated with a couple of weeks of antibiotics—certainly more than a day but less than a lifetime. If streptococci remain in the throat, there is a chance of relapse and of more serious sequelae, such as rheumatic fever or kidney disease. If any *B. burgdorferi* manage to survive antibiotic therapy for early Lyme disease, could late Lyme disease develop in the future? It depends how "survive" is defined. If the spirochetes are not really alive in the way we usually define, could the cellular residue provoke a harmful reaction? This discussion continues in the next chapter.

Are Two Antibiotics Better than One?

One justification for using more than one antibiotic at a time is to prevent the emergence of antibiotic-resistant bacteria during treatment.
This strategy is routinely used in treating people who have tuberculosis
(or, for antiviral drugs, HIV infection). Someone with an active case of
tuberculosis may have huge numbers of bacteria in the lungs. Chances
are that at least one member of the bacterial population is already resistant to an antibiotic. If only one antibiotic is used, the susceptible bacteria will die, but the resistant mutant bacterium will grow in number to
two, four, eight, and so on until it has supplanted the dying antibiotic-
susceptible cells. The patient is just as sick but now has a bacterial population that cannot be wiped out with that particular antibiotic. When two
or more antibiotics are used in combination, however, one of the antibiotics will kill a mutant that is resistant to the other—if the two antibiotics work in different ways. The odds that a single bacterium would
become resistant to two or more different antibiotics at the same time
are very low.

Is this experience with tuberculosis relevant to Lyme disease? Probably not. Mutants of *B. burgdorferi* that are resistant to standard therapies have not been detected in the environment or in patients. Different
strains of these bacteria from various locales in the world vary to a small
degree in their susceptibilities to some antibiotics in the laboratory, but
this amount of variation is not worrisome. Truly resistant bacteria usually differ from their susceptible counterparts by a large margin. And,
as we saw in the last chapter, even if a mutant arose in an infected individual, it would be highly unlikely to be passed beyond a single patient
the way a newly resistant tuberculosis bacillus could be.

A second justification for choosing two antibiotics over one is to take
advantage of their different modes of action to achieve with the combination a substantially greater effect than could be achieved with either
individually. A theory behind this strategy might claim that an antibiotic
like amoxicillin is fine for killing off 90 to 99 percent of the spirochetes,

but for what remains, which may not be replicating, the addition of a second antibiotic that works in a different way from a penicillin-type drug could target these persisters. One caution about taking this approach is that two antibiotics may actually interfere with each other in their actions on the bacteria, so the outcome ends up worse than with either drug by itself. There have been some well-designed clinical trials of sequentially administered antibiotics, such as intravenous ceftriaxone followed by oral doxycycline, but there are not published reports of blinded, prospective controlled trials of two or more drugs at the same time.

Should Asymptomatic Infection Be Treated?

The phrase "asymptomatic infection" sounds like an oxymoron. Most people would assume that infections make people sick. That's not always true, and sticklers make a good argument that "asymptomatic" should mean exactly that: no symptoms. But that's a pretty high bar to clear when querying people about their health months to years prior to the interview or questionnaire. So "asymptomatic" probably includes some past instances of a brief illness that was either too mild to come to medical attention or so undifferentiated as to be indistinguishable from an inconsequential viral infection. By either definition, asymptomatic infection seems to be less common in North America than in Europe, where they have some less invasive *Borrelia* types. But whether truly asymptomatic or not, people do have Lyme disease tests who have no recollection of having Lyme disease and don't now have symptoms of infection. If that test turns out positive by both EIA and Western blot, what to do? Let's assume that this was a true-positive, that is, the patient truly had *B. burgdorferi* infection and made antibodies in response. Are these seropositive people at risk of developing late Lyme disease in the future?

There are few data germane to the long-term outcome for seropositive asymptomatic persons. Since the early 1980s, most people with symptomatic Lyme disease in the United States have been treated with antibiotics. (The prognosis for the treated group is a topic of the next chapter.)

A community survey with two hundred subjects in the early 1980s on a New York island found that about 10 percent had antibodies to *B. burgdorferi*; less than half of these 10 percent had a history of Lyme disease.[1] But what happened to the seropositive individuals without Lyme disease over the subsequent years is not recorded. There was a short window in time when Lyme disease patients were studied but not treated with antibiotics. A follow-up study of some of those patients found that those who had untreated Bell's palsy tended to have more chronic symptoms and disability than did treated Bell's palsy patients.[2] But, by any measure, Bell's palsy would hardly count as "asymptomatic" or be easily overlooked; it would likely be treated these days in areas where Lyme disease occurs.

Some pertinent studies are from Europe, which, as stated, may have a higher frequency of inapparent infections than North America. A survey over three years of several hundred off-trail runners in Switzerland's forests found sixteen whose serum samples turned from negative to positive during the study period; only two of the sixteen developed symptoms of Lyme disease during the follow-up period.[3] A prospective study of forestry workers in the Netherlands found that about 28 percent had antibodies to *B. burgdorferi* at the start.[4] Some of these seropositive workers had had Lyme disease, but thirty-two were seropositive without a history of Lyme disease and were asymptomatic at the first exam. Twelve months later, none of the thirty-two had become symptomatic.

The follow-up periods for these studies were about a year. What are the prospects further out for untreated seropositive individuals? The worry is that Lyme disease will be like tuberculosis or syphilis. Initial infections with these pathogens produce an early illness with symptoms and signs, such as syphilitic skin lesion on the body, but then the microbes retreat under pressure from the immune system. They lie dormant but years later may emerge again to cause disease, often more severe than the infection's first incarnation. But something like this does not seem to occur with Lyme disease *Borrelia* infections. The analogy of Lyme disease with syphilis in terms of infection stages only goes so far. If early infection resolves on its own without treatment, there is a definable risk of late infection, particularly of the nervous system or joints, appearing during the first year after a symptom-free interval. But this risk declines after the first few months to the point that true Lyme disease appearing again two or more years after an earlier episode is much more likely a new infection than a reactivation of the first one. So those of us walking around with long-standing antibodies to *B. burgdorferi* and feeling otherwise fine are not bearing a ticking bomb, timed to go off years hence. Although there are strong arguments for treating asymptomatic individuals who are seropositive for syphilis, there is not as compelling a rationale for antibiotic intervention to prevent long-delayed activation of Lyme disease.

Antibiotic Therapy for Pets

The risk of *B. burgdorferi* infection for dogs exceeds the risk for people in a given geographic area. More than half the dogs may be antibody-positive in some high-risk areas, while the antibody-positive rate for human residents in the same area is around 10 percent.[5] But exposed dogs are less likely than humans to become sick after infection.[6] What disease occurs in dogs is mainly in puppies and young dogs and in the form of arthritis leading to lameness several weeks after an infecting tick bite. There are reports of an uncommon but serious kidney disorder that is suspected to be caused by a combination of *B. burgdorferi* infection itself and the immune response to it. The most commonly used diagnostic test for antibodies in dogs is an EIA test with C6 peptide, which is similar to the human assay. A follow-up Western blot is seldom performed, so there may be more false-positive test results for dogs.

For seropositive dogs with joint disease or the kidney disease, antibiotics are commonly prescribed. If *B. burgdorferi* infection was the cause, the lameness tends to respond more quickly to antibiotics than does the longer-standing kidney disorder. Although there have been studies of antibiotics in experimental infections of dogs, there still have not been any controlled clinical trials for naturally acquired infections in the community. Doxycycline by mouth is the most common therapy. Unlike amoxicillin or ceftriaxone, this tetracycline antibiotic is also a suitable treatment of *Anaplasma phagocytophilum* (chapter 11), another pathogen carried by deer ticks. Alternative oral antibiotics are the macrolide azithromycin or the penicillin amoxicillin. If an intravenous antibiotic is required, ceftriaxone is the primary option.

Should healthy dogs be screened for *B. burgdorferi* antibodies? There is controversy about this, with most arguments against this practice. A prospective study of at-risk outdoor Bernese mountain dogs in Switzerland looked at seropositive animals that showed no signs of illness at the time of the blood draw.[7] When they were examined two to three years later, they were still free of lameness and kidney disease. So, as for humans, the seropositive state in the absence of illness or organ dysfunction is not, by itself, an indication for treatment.

Some cats may be at as great a risk of *B. burgdorferi* infection as dogs in areas with Lyme disease. But in comparison to dogs, there is much less information on either experimental or natural infections in cats. If Lyme disease is suspected in a cat, the usual antibiotic for treatment is doxycycline.

Other Treatments: Conventional and Unconventional

Unorthodox theories about therapies for Lyme disease abound. The claimants would either replace standard antibiotic treatments for Lyme disease or supplement them. Some have been downright dangerous, such as the scheme for curing "chronic Lyme disease" by purposefully infecting patients with malaria parasites in a clinic in Mexico (malariotherapy).* Others have advocated treatments with silver salts, usually in the form of "colloidal silver," or "chelation therapy" to remove heavy metals from the blood—which would seem at cross-purpose to a silver-based treatment! Some operators of hyperbaric oxygen chamber facilities offer treatments with the purported aim of either killing the

---

*The malaria-based treatment idea traces back to the now-discredited "fever therapy" for syphilis of the nervous system, which—to the current prize committee's likely chagrin—won its originator the Nobel Prize for Medicine in 1927.

oxygen-sensitive spirochetes or revving up the immune system, but without published scientific backing to support use of this expensive procedure. The examples could go on (and are detailed at the Quackwatch website's Lyme disease page, which is listed in the "Trusted Internet Sites").

Of course, management of Lyme disease and its manifestations, especially in its late forms, does not rest solely on antibiotics. A patient who has Lyme disease with invasion of the heart and a dangerously slow heart rate would likely receive a temporary pacemaker while the antibiotic takes effect and the natural rhythm restores. Someone with chronic Lyme arthritis that responds poorly at first to antibiotics may benefit from a nonsteroidal anti-inflammatory drug, like ibuprofen or naproxen, or something more specific for arthritis, like hydroxychloroquine, and ultimately may need minor surgery, a synovectomy, to remove inflamed joint tissue.

Prednisone and related corticosteroid medicines at one time constituted the only treatment for some cases of Lyme arthritis or neurologic disease, and some people seemed to respond to these agents alone. When antibiotics were discovered to be so beneficial for people with Lyme disease, however, corticosteroids were relegated to an occasional supplementary role. Now they are sometimes used to treat people with heart involvement of Lyme disease, to speed return of their abnormal heartbeat to normal. Some physicians also prescribe a drug such as prednisone to relieve the pains and dysfunction of nervous system involvement by Lyme disease *Borrelia*, especially in Europe. The reasoning behind this approach to treatment is that the inflammation itself is producing damage, and that corticosteroids can limit this damage by reducing swelling and the number of cells in a critical area such as the heart, brain, or spinal cord. The other side of the coin, however, is that corticosteroids will also suppress the patient's immune response to some extent. A corticosteroid may impair a person's own defenses against the spirochete. Experiments with dogs showed that a long-dormant *B. burgdorferi* infection could reactivate when the dogs are put on a corticosteroid.[8] That is why, if corticosteroids are used, they are used for only a brief period, and only in conjunction with antibiotic treatment.

Treatment Recommendations

The consensus recommendations summarized below are current at the time of writing. I expect further modifications of treatment protocols in the future. The identity of the infecting strain of *B. burgdorferi* may come to play a role in the duration of therapy, with infection by a more invasive strain calling for a more extended course of treatment. Although the pipeline flow for new antibiotics from the pharmaceutical industry had slowed to a trickle, the alarming increase in antibiotic resistance in various medically important bacteria, like staphylococci, has opened the research and development valves again. With the explosion of research and applications for routine genome sequencing for individuals as well as "personalized medicine," there will likely be further advances in identifying those who are at greater risk of an adverse effect from a drug or who require a higher dose of the medicine for it to be fully effective. And finally, we should have a better sense of the significance of persisting spirochete cells or parts thereof, as discussed in the next chapter, for treatment decisions.

A good starting place for up-to-date treatment recommendations is one of the clinical practice guidelines put out by nonprofit organizations of primary care physicians, like pediatricians, or medical specialists, like neurologists.* Often these consensus recommendations are vetted and adopted by government public health agencies. These organizations and professional societies exist in both North America and Europe. One that has gained the most attention with regard to Lyme disease is the Infectious Diseases Society of America (IDSA), which has a membership of about 10,000 infectious disease physicians and scientists.† The IDSA guidelines are representative of the "Earth" view of Lyme disease (as defined in the introduction). The original IDSA recommendations appeared

---

* Guidelines are becoming widespread in medical care. For better or worse, they lead to a greater standardization of care, as physicians' performances or malpractice suits may be evaluated in the context of adherence to or deviance from consensus guidelines.

† I am a long-standing member of the IDSA but did not participate in the writing of former or current guidelines for Lyme disease.

in 2000 and then in revised form in 2006.[9] They were reviewed by an independent expert panel in 2010, which in the end agreed with all recommendations.[10] The IDSA recommendations are largely in agreement with those of American Academy of Neurology,[11] as well as the following organizations outside the United States: Canadian Public Health Network, European Federation of Neurological Societies, European Union of Concerted Action on Lyme Borreliosis, and German Society for Hygiene and Microbiology.

What may be confusing—and eventually frustrating—for those not familiar with the thirty-year history of controversy is that there is a second set of Lyme disease guidelines, which at many points are at odds with those of the IDSA and allied societies. These are from the International Lyme and Associated Diseases Society (ILADS).[12] In my idiosyncratic conception, the ILADS guidelines are representative of the "Twin Earth" view. In other words, there is such a fundamental difference in the definitions of Lyme disease that the two sets of guidelines are incommensurate on many issues, such as treatment. The ILADS "Lyme disease" tent is bigger than this book would allow for and admits some diagnoses lacking customary evidence of *B. burgdorferi* infection. The idiom about "comparing apples and oranges" comes to mind. No doubt these disabling, persisting illnesses—sometimes labeled "chronic Lyme disease" but outside my conception of *B. burgdorferi* infection—deserve attention and our better understanding. However, that would be beyond the limits of this book and the capacity of its author.

With these caveats to the fore, are there some general principles for treatment that may stand the test of time? While acknowledging that exceptions are inevitable, we can describe treatment this way:

- Oral antibiotics are used in the treatment of early localized infection, early disseminated infection, and late infection, with some exceptions (see below).* The first-line drugs are chosen from tetracyclines (such as doxycycline), penicillins (such as amoxicillin), and cephalosporins (such as cefuroxime). Currently

---

* Antibiotic treatment after a tick bite or exposure is considered in chapter 13.

available macrolides (such as azithromycin) are second-line drugs.

- Antibiotics that are effective if taken just once, twice, or at most three times a day are preferred for patient compliance. This more convenient dosage schedule may be at the cost, though, of a needlessly broad "spectrum" of antibiotic activity, which may lead to more drastic alterations of the intestinal microbiome and then diarrhea as a consequence.

- Intravenous cephalosporin or penicillin antibiotics are indicated for patients with heart involvement and a dangerous slowing of the rhythm and for those with neurologic disorder that is consistent with invasion of the central nervous system (CNS) by the spirochete. This includes signs of meningitis and brain dysfunction, but not necessarily Bell's (facial nerve) palsy or nerve involvement alone. To accurately assess CNS involvement, it may be necessary to do a lumbar puncture (spinal tap) and examine the cerebrospinal fluid.

- The arthritis of late Lyme disease was once routinely treated with intravenous antibiotics, but now intravenous antibiotics are recommended only if an adequate trial (up to 4 weeks) of an oral antibiotic has failed or if the arthritis is accompanied by signs of CNS involvement.

- There is little evidence to support a treatment duration of more than 21 days for most cases of Lyme disease. The usual course is 14 days. As few as 10 days may be sufficient for the majority of patients, but 7 days of any oral drug is probably inadequate. (There may in the future be modifications tailored for individual patients as we learn more about the outcomes of infections with more invasive strains and about variations in how different individuals' immune systems inherently respond to *Borrelia* infection.)

- Three- to four-week-long therapies may be justified for cases of Lyme arthritis, Bell's palsy, or heart involvement.

- There are many possible contraindications to the different antibiotics, including a history of a true allergic reaction, but an overriding warning is against use of a tetracycline class antibiotic during pregnancy and breast-feeding, or for children younger than 9 years.

■ If the skin infection called cellulitis caused by staphylococci or streptococci is a possible diagnosis (that is, the person's skin condition may not be erythema migrans for certain), then either cefuroxime or amoxicillin in combination with clavulanate (a compound that blocks an antibiotic-resistance mechanism of staphylococci) gets the nod over doxycycline.

■ If the illness may be human granulocytic anaplasmosis instead of or in addition to Lyme disease (chapter 11), then doxycycline (or other drug with equivalent activity against bacteria proliferating inside cells) is preferred.

Remember, guidelines are just that: guides, not rules. If the guidelines are evidence based and thoughtfully drawn up by panels of experienced experts, they can be useful roadmaps to achieving high-quality care for cases that accurately and comfortably align with the guidelines' target. But a patient's illness may not fit so cleanly into one or another category, or the patient's illness may be a toss-up between two diagnoses (or three). This is when the art as well as the science of medicine is called for.

Treatment: A Personal View

In my conversations and correspondences over the years with various people about their options for treating Lyme disease, I have noticed that many have a different general conception of therapy than mine. Their expectations for treatment are more in line with those for a noninfectious disease. They may be seeking or have already found a doctor with special knowledge and skill, often denoted by the code words "Lyme literate." The person is essentially asking about the next level up for treatment of what has been labeled Lyme disease, now that either the standard regimen or one of the alternative methods has failed. Their view is of a graded series of treatment methods, with concomitantly higher levels of riskiness, unpredictability, and/or financial cost, as a person progresses from one to another. This view may correspond pretty closely to reality for some serious noninfectious diseases such as cancer, autoimmune conditions such as rheumatoid arthritis, or psychiatric disorders such as major depression. If the first line of therapy brings no or only partial relief, a second-line treatment protocol is tried. Often the second-line (or third or fourth) therapy has a higher likelihood of adverse effects, but this is justified if the inherently safer first-line treatment comes up short. For instance, a patient with cancer may go through a succession of different types of chemotherapies or therapeutic antibody treatments. Eventually the patient may have a bone marrow transplant or some other major procedure.

My view is that the treatment options for true *B. burgdorferi* infection are more limited. For the majority of cases, there are oral antibiotics, which might be given for as long as a month but usually for half that time, and seldom for longer. Less commonly, an intravenous antibiotic is administered, but again, rarely for more than a month. If a month of intravenous ceftriaxone has provided little or no benefit for a patient with a diagnosis of Lyme disease, another round of antibiotics is not likely to help. And there is no equivalent to, for example, infusions of a precious therapeutic antibody, electroshock therapy, ion-beam radiotherapy, or a bone marrow transplant at the end of a list of treatment options that offer hope of a remission if not a cure. At this point in the patient's treatment, the diagnosis of active *B. burgdorferi* infection may need to be questioned. As we will see in the next chapter, there may be other explanations for continued or recurrent symptoms after the antibiotic ends.

Chapter 10
After Antibiotic Therapy Ends

■ **Case L.** In the month of June, a previously healthy 30-year-old man living in Massachusetts developed an expanding red rash on one ankle accompanied by fever, loss of appetite, and fatigue. Without any treatment, these symptoms resolved within 5 days. In December he reported persistent swelling of the left knee. A laboratory test showed that his blood had antibodies to *Borrelia burgdorferi*. On the basis of the medical history, the physical exam, and the lab results, a diagnosis of Lyme disease was made, and the patient was treated with an intravenous antibiotic for two weeks. The knee swelling resolved over the next three months, but he continued to have fatigue and diffuse pain and stiffness in the wrists, elbows, shoulders, and knees. Because of the persistent symptoms, he was given another course of the intravenous antibiotic a year later. After an initial improvement, his fatigue, stiffness, and joint achiness worsened again over the next several months, and these symptoms were accompanied by headaches, memory difficulties, irritability, and early morning awakening. On physical exam, his joints were normal, but he had tenderness to deep touch at various places on his body, such as the back of the neck and the juncture of arm and shoulder.[1]

I'll be frank at the outset, to head off disappointment later. This chapter concerns persisting symptoms after completion of conventional antibiotic treatment for *B. burgdorferi* infection. "Conventional" means use of one of the recommended penicillins, tetracyclines, cephalosporins, or macrolides and lengths of therapy in the 10-day to 21-day range (chapter 9). There might have been a second round of antibiotics of similar duration. "Infection" means that both of the following conditions are met: (1) the patient had a plausible risk of acquiring the infection, by virtue of exposure to deer ticks at locations where and when transmission was feasible (chapter 4); and (2) the illness after the exposure was compatible in its symptoms, signs, and sequence over time with Lyme disease. Preferably, antibody testing supports the clinical diagnosis, but for justifiable reasons that may not have been performed for typical erythema migrans (chapter 1). The initial illness of Case L, which is summarized above, meets these conditions. The features need not check every box of CDC's Lyme disease–reporting criteria, which for the sake of public health monitoring is stringent. But the features ought to put it in the same ball park as Lyme disease and not a ball park in another city. For those with illnesses that have slim chance of being the consequence of *B. burgdorferi*,* much of what follows may unhappily be of little help or comfort.

Case L's is not the usual outcome—the great majority of patients recover with treatment or shortly thereafter—but neither is a story like L's that rare. Enough patients continue to have symptoms after completion of therapy to spawn and feed the controversy about how long treatment should continue. A lot of names, including "chronic Lyme disease," "chronic Lyme borreliosis," or simply "chronic Lyme," have been applied to an illness like what Case L experienced. The term I adopt is "posttreatment Lyme disease syndrome," or PTLDS,[2] because it explicitly specifies previous therapy as a criterion. This choice was made over the less specific term "post–Lyme disease syndrome."

This chapter looks at the frequency of PTLDS, and then relates that condition to two other enigmatic disorders, chronic fatigue syndrome and fibromyalgia, with which it seems to overlap. I next consider the evidence on whether further antibiotic therapy is of benefit or not for

---

*What's an example of a "slim chance"? Long-standing or recurrent symptoms of a nonspecific nature, like generalized achiness, difficulty concentrating, and fatigue, in a resident of Arizona who has not traveled out of the state.

PTLDS, and then end the chapter with a discussion about the possible explanations for PTLDS and a look back at an infection that was in its day as notorious as Lyme disease is now.

Post-Treatment Lyme Disease Syndrome

The last two chapters reported the effectiveness of certain antibiotics against *B. burgdorferi* in different situations: the laboratory, experimental animals, and clinical trials with humans. These research findings back up the collective and cumulative experience of hundreds of physicians: there is faster resolution of infection and a lower risk of late Lyme disease with antibiotic treatment. Other studies that followed up on patients after episodes of Lyme disease have likewise found evidence of antibiotic therapy's benefit in dramatically reducing occurrences of late disease, like arthritis and neurologic disorders. The large majority of people who have had documented Lyme disease and received adequate therapy are indistinguishable from controls of the same age and sex in terms of their perceived health at a checkup several years later.[3]

If that was the whole story, the book could end here. But it's not. In these trials and in other outcome studies, between 5 to 10 percent of the subjects who received an antibiotic continued to have some symptoms for six months or longer after treatment ended.[4] The rash may be just a memory, the heart is beating regularly again, the joints are once again limber, but everything is still not right. Their overall health is better than it was while suffering Lyme disease, but they may continue to lack energy, become easily fatigued, and have recurrent muscle and joint aches and headaches. Some complain of mental sluggishness and difficulty concentrating. A study of people with erythema migrans who were followed prospectively from the time of diagnosis and treatment revealed that about one-third of them reported new-onset fatigue, aches and pains, and/or concentration and memory problems at the six-month followup.[5] These persistent symptoms are distinguishable from a second (or third) episode of *B. burgdorferi* infection, which might manifest as a new erythema migrans rash.

This phenomenon is not unique to Lyme disease. A postinfection syndrome is a loose collection of symptoms following a documented or suspected infection.[6] Fatigue is a common feature. People experience similar debilitating symptoms as an aftermath of many common types of infection, including influenza, hepatitis, and infectious mononucleosis. A sampling of other, less well-known infections for which this has been reported includes the following: (1) the intestinal parasite disease giardiasis, which occurs globally; (2) mosquito-borne chikungunya virus infection; (3) Ross River virus infection in Australia and the South Pacific; and (4) the bacterial zoonosis Q fever, common in sheep-raising areas. In these instances of a postinfection disorder, the infectious agent presumably is either eliminated or no longer multiplying. Although the postinfection syndromes are associated with many different types of microbes—from viruses to bacteria to one-celled animals—they have more similarities than differences.[7] Curiously, these postinfection syndromes also have similarities and overlaps with two other perplexing but not uncommon disorders: chronic fatigue syndrome and fibromyalgia. Case L had features of both PTLDS and fibromyalgia.

Chronic Fatigue Syndrome and Fibromyalgia

If you were to look in the table of contents of an old medical textbook, you would not find a listing for fibromyalgia or chronic fatigue syndrome. That's not to say these conditions are new under the sun, however. These older textbooks described other conditions, such as "fibrositis," which had many of the features of fibromyalgia, and "neurasthenia," whose symptoms were similar to those of what is today called chronic fatigue syndrome (CFS). At the turn of the previous century, neurasthenia, commonly considered a disease of "brain workers" and the consequence of "nervous exhaustion," affected several famous or soon-to-be-famous writers, artists, politicians, and other leaders of the day.

Like both neurasthenia and fibrositis, fibromyalgia and CFS are located close to the secondary quality (chapter 5), or subjective pole, of the spectrum of diagnosis. The patient's symptoms or their responses to the physician's questions form the principal basis for diagnosis. Because fibromyalgia and CFS symptoms overlap those for PTLDS, they may be confused with that disorder, as well as with each other. Within the overlapping definitions of CFS, fibromyalgia, and PTLDS are several shared symptoms: fatigue, aches and pains of the musculoskeletal system, headaches, sleep disturbances, and a perception of diminished mental function (impairment of short-term memory or concentration). In Europe, the equivalent term for CFS is the tongue twister myalgic encephalomyelitis, which at least gives hints of the muscle aches (myalgic), effects on the brain (encephalo-) and spinal cord (-myel-), and an inflammatory component (-itis).

The major emphasis for the diagnosis of CFS is on profound fatigue that is not relieved by bed rest, has persisted for at least six months, and has substantially reduced or impaired the patient's average daily activity. Other symptoms that are more suggestive of CFS than the other disorders considered here are a recurring sore throat, tender lymph nodes in the neck or under the arms, and fatigue that is made worse by physical or mental exertion. The diagnosis of CFS assumes that other diseases, such as a thyroid condition, liver disease, or anemia, which can produce fatigue, have been excluded by the patient's medical history, a physical examination, or laboratory tests. No specific diagnostic tests are available for CFS.

The more distinctive aspect of fibromyalgia is pain, both that experienced by the patient as a symptom and that elicited by the physician on physical examination. The pain may seem to the patient to be centered in the joints, but the greatest involvement is actually of the musculoskeletal tissues outside joints—hence the specification of fibrous tissue (fibro-) as well as painful muscles. Fibromyalgia is one of the "nonarticular rheumatism" conditions; another is bursitis. One of the principal phenomena of fibromyalgia is a heightened perception of pain from a stimulus that would not normally be considered painful. In other words, the person with fibromyalgia feels discomfort under conditions that most people would not view as uncomfortable.

The musculoskeletal pains of fibromyalgia are usually worst in the morning and are accompanied by stiffness, intensified by changes in the weather, and temporarily relieved by heat. In examining the patient, the physician finds specific and well-localized points of tenderness and pain when pressure is applied. The patient's responses to palpation and the number and location of the tender points are important factors in making the diagnosis. Case L was reported in a medical journal as an instance of fibromyalgia following Lyme disease. But if Case L had not had the signs of out-of-the-ordinary tenderness to deep touch, his condition would simply have been called post-treatment Lyme disease syndrome.

Like a diagnosis of chronic fatigue syndrome, a diagnosis of fibromyalgia depends on the exclusion of other conditions that could produce generalized musculoskeletal pain of this sort as well as fatigue and sleep disturbances. The laboratory studies that are performed to rule out other diseases do not confirm a diagnosis of fibromyalgia. To date there are no laboratory tests, singly or in combination, that prove the existence of this condition.

Although CFS and fibromyalgia can arise insidiously without prior illness, many cases of CFS and fibromyalgia have been associated in time with a preceding infection. When there is a clear relationship between the onset of symptoms and a recent infection, CFS and fibromyalgia can be considered postinfection syndromes, like PTLDS, and the same questions about mechanisms apply.

PTLDS Treatment Trials

In my book on Lyme disease published in 1996, I commented on claims about long-term antibiotic treatment for persistent symptoms: "A collection of anecdotal accounts, as dramatic as they may sound, and as sincerely as they may be presented, is of limited value in the end. What these accounts of therapeutic success prompt is not a blanket change in treatment recommendations but controlled trials to see if all or part of the theory holds up."[8]

The years since have witnessed progress. From 1997 to 2005, there were four blinded, placebo-controlled trials, all sponsored by the National

Institutes of Health and mostly in the northeastern United States.[9-12] Though they differed in several respects, the trials had in common this overarching question: Does intravenous antibiotic therapy for one to two months provide greater benefit than harm for patients with persistent symptoms and disability after documented or suspected Lyme disease? The results are in, and they have been analyzed. Are the results definitive, settling the issues for good? Of course not. Even the best medical research can seldom make such a claim. One could debate on the postgame show for these trials whether or not there was some evidence of partial benefit in this study or that.[13,14] But the bald facts are that there was no obvious sustained benefit from receiving ceftriaxone intravenously every day for up to ten weeks or receiving ceftriaxone for one month and then oral doxycycline for another two months. In statistical parlance, the null hypothesis of equivalence of antibiotic and placebo in effect could not be rejected. Certainly, when weighed against the medical risks of prolonged intravenous therapy—not to mention the high monetary costs for such treatments—the argument favoring long courses of intravenous therapy came up short.

Nevertheless, these studies leave room for reasonable speculation about minor effects or benefits on the margins. The studies' authors cannot conclude that there is *no* benefit from antibiotics, only that if benefit was present, it was not great enough to be discernible in groups of these sizes. The number of subjects was comparatively small, averaging only thirty or so patients with suspected Lyme disease in each study. There were simply not enough treatment and placebo patients in these studies to rule out either beneficial effect in a subset of the patients or a modest effect across all patients. Each study had different criteria for enrollment of subjects, but all the trials struggled to enroll enough subjects who would both meet the case definition and agree to participate. Less than 5 percent of the individuals screened for these trials got to the point where they were randomly assigned to treatment or placebo arms. Extrapolating the results of these studies to circumstances beyond the case definitions is possible but is best done cautiously.

From that contrarian stance, I return to the question of whether a minor benefit, even if provable in a larger study, is worth a risk upwards of

20 percent of a life-threatening complication, such as in Case K of chapter 8, or other serious adverse events that the investigators observed among those on lengthy intravenous therapies. Would it be ethical even to carry out a larger study knowing the risks subjects would be taking? A rejoinder might be, "Well, these studies and the risks of further intravenous studies settle the intravenous ceftriaxone issue, but what about placebo-controlled trials of long-term oral therapy for PTLDS?" To that, the response is, "We'll have to wait." Only one small trial of oral doxycycline in Europe has been reported to date.[15] It showed no difference versus placebo from three weeks of treatment, but that is unlikely to be the last word. Stay tuned.

Unintended Effects of Antibiotics

The discouraging results of controlled trials of an intravenous antibiotic notwithstanding, a continuing puzzle is that some patients in uncontrolled studies continue to report feeling better while on antibiotic therapy.[16] Their testimonials on this point populate Lyme disease–oriented websites, blogs, and social media. What distresses them and their caregivers is the return or continuation of symptoms after what should have been an adequate length of treatment. Antibiotics are again started, with partial or complete improvement. The drugs may be taken conscientiously for weeks or even months, but, in what is not an unusual story, the symptoms return sooner or later after the antibiotics are stopped. Let's accept at face value these experiences and, for the sake of argument, assume that the placebo effect does not account for all of them (that is, people don't feel better just because they are taking a pill that they think will make them feel better). Could a penicillin, cephalosporin, tetracycline, or macrolide be having an action in patients that is not attributable to their antibacterial activities?*

---

*This leaves aside the effect of antibiotics on the intestinal bacteria of our microbiomes, which could in turn have systemic consequences. But it is hard to imagine that antibiotics would have a salutary effect through this mechanism.

A drug's side effects are its unlovable aspect. Side effects are to be endured or, if bad enough, to be grounds for stopping the medicine. But could some of these adverse effects be telling us something positive? Why does nausea occur for some people taking a macrolide, dizziness with a tetracycline, or nervous excitement with high doses of a penicillin? Could there be other activities of the antibiotic that are either separate from or part of their antibacterial action? An early recognition of possible benefits from these "non-antimicrobial" functions of antibiotics was in the treatment of common acne and the related condition rosacea.[17] Theories of acne treatment were once based solely on the notion that antibiotics would inhibit certain skin bacteria from growing, thereby preventing pimples. Although to a certain extent these antibiotics do change the bacterial flora of the skin, perhaps a more significant action of the antibiotics is reduction of inflammation. Tetracycline and related antibiotics are some of the most frequently prescribed medicines for acne; teenagers may take tetracycline every day for months at a time.

Antibiotics are not on Earth solely for people's welfare as handy medicines—if we are clever enough to discover them. As chapter 8 described, the original antibiotics were potent chemical weapons made by bacteria and fungi against their competitors among other species in their environments. Although humans have adopted and modified them for good purposes, antibiotics probably predated multicellular life on Earth. But the story goes back even further. A bacterial cell's machinery to make an antibiotic is complex, involving several steps that are tightly coordinated. A molecule with antibacterial properties would have more likely arisen from modifications of molecules already in existence rather than been "invented" de novo. Evolution science abounds with examples of a repurposing of an existing molecule through a series of slight alterations.

What might the original products of this cell machinery have been? Antibiotics probably derive from small molecules that have signaling activities. In other words, they are chemical means of communication, informing other cells in the vicinity of conditions the secreting cell is sensing. An example is the release of one of these signaling molecules when the density of bacteria in a given space reaches a certain level. The bacteria may interact in a different way after this alert goes out. One of these ways is to gather into a film of bacteria on a surface rather than swim free on their own in the liquid environment. Experimental evidence supporting this hypothesis about antibiotic origins is a large set of research studies showing that various antibiotics at low concentrations—below that at which bacteria are inhibited or killed—will affect the formation of these films.[18]

Along with science's recognition of non-antimicrobial functions of antibiotics is pharmaceutical companies' practice of recycling currently approved drugs for other purposes. There is a considerable advantage over starting from scratch in finding new uses (and new patents) for a medicine that has already gone through the very costly development and clinical testing process. Several investigations along these lines with antibiotics are under way. What is called "subantimicrobial dose doxycycline," under the brand name Periostat, has been approved by the FDA and other regulatory agencies in Canada and Europe as part of the therapy for chronic periodontal disease. This and other tetracyclines, among their other actions, suppress an enzyme that has a prominent role in the tissue damage that can occur with inflammation. It does this at a dose that is only one-fifth as large as what is used to treat Lyme disease and that causes fewer side effects.

Low-dose macrolides are used as long-term medicines for people who have cystic fibrosis, because some of them can inhibit formation in the lungs of bacteria's tenacious film. The antibiotic achieves this even with bacteria that are resistant to its growth-inhibiting property. The macrolides are also thought to have anti-inflammatory actions that are apart from their effects on bacteria.[19] Even cephalosporins, which along with penicillins were the antibiotics that seemed most restricted in their actions on bacteria, can directly affect the levels of an important brain chemical, glutamate.[20] High levels of glutamate in certain areas of the brain cause excitation of neurons and damage. Ceftriaxone in particular, because of its convenient once-a-day dosing, is being investigated in various neurologic settings, such as for amyotrophic lateral sclerosis (also known as Lou Gehrig's disease) and for brain trauma and stroke. In one of the controlled trials of long-course intravenous ceftriaxone for suspected Lyme disease of the brain, there was some improvement in mental or cognitive functioning in the ceftriaxone group compared with the placebo group, but only while on the antibiotic or just after and not a few months later.

How pertinent are these non-antimicrobial activities of antibiotics to PTLDS or, for that matter, the broader group of postinfection syndromes, fibromyalgia, chronic fatigue syndrome, and "chronic Lyme disease"? That remains a question. As I see it, we cannot rule out that some of the alleged benefits of antibiotics are attributable not to the antibacterial function of these drugs but to their anti-inflammatory or neuron-protective effects.

What Causes PTLDS?

Perhaps because our view of infectious disease is shaped by an orientation toward the importance of vaccinations, chemotherapeutic treatment, and supportive therapy, we tend to overlook the fact that animals and people have been exposed to, and have survived, the effects of disease-causing organisms through millions of years of evolutionary history.[21]

Benjamin Hart, 1988

Infected crickets become lethargic and their characteristic chirps slow in tempo; they are less receptive to sex. Infected mammals have not only fever but also diminished appetite, sleepiness, and mental depression. Think of the cat that retires to the back of the closet when ill. These are examples of "sickness behavior," which from an evolutionary perspective can be counted as adaptive—that is, in the better interests of the species. The elevated temperature can slow the pathogen's growth rate; the lower intake of food may starve it of needed nutrients, like iron; the reduced physical and mental activity conserves energy for the first priority: fighting off the infection.

Cytokines and nervous system chemicals that increase in varieties and amounts during infection can account for the sickness behavior, but they may have enduring effects. When some cytokines are used as drugs themselves and given to people as treatments for cancer and other conditions, the patients experience fatigue, muscle aches, and difficulty concentrating. Such symptoms are similar to the symptoms experienced by people with PTLDS.

What follows from these observations is this conjecture about PTLDS and other postinfection syndromes: it is an inappropriate extension of this otherwise beneficial host response and involves immune, endocrine, and nervous systems. The anti-infection machine has been switched on, so to speak, but the usually reliable automatic turn-off switch does not function at the right time, and the machine continues to operate. What is of benefit for a human or other animal fighting an infection, such as the urge to rest in order to recuperate, becomes over time the symptom of easy fatigability—and this can become quite a liability if a person has a family and a job to attend to.

The preinfection mood or temperament of the patient may also have a bearing on the time required for convalescence. The effect of a person's mental state on infection outcome was studied in two situations. One was a college in which students had taken a personality test at the start of the year. Then those who came down with infectious mononucleosis were examined and further questioned.[22] In the second study, an epidemic of influenza was tracked in a group of people who had taken psychological tests just before the epidemic hit.[23] In both studies it was found that people with a tendency toward depression and those who had a less optimistic outlook took longer to convalesce from infection than others.

These two concepts about postinfection syndromes—a runaway host response and psychological vulnerability—in their purest expressions assume that the infection is over and all traces of the microbe are gone. But, as those advocating long-term therapy argue,[24] the infection may continue in a stealth form. According to this view, the spirochetes remain viable and cause mischief, while evading the threat from standard courses of an antibiotic. The clinical trials of prolonged antibiotic therapy for PTLDS indirectly addressed this possibility. If the spirochetes remained alive, there was no evidence that they were affected by up to three months of continuous therapy. One might invoke for comparison the infection tuberculosis, which requires months-long treatments for cure. But it would be unusual if there was no discernible improvement in a patient's health status within one month of starting a regimen for tuberculosis. In studies of experimental infections of mice and dogs, appropriately dosed antibiotics resulted in a large drop in the number of spirochetes recovered in culture from a variety of tissues, including skin and deeper organs.[25] If a positive test for antibody to *B. burgdorferi* is analogous to finding a smoking gun in a police procedural, then isolation of the spirochete in culture is like finding in the victim a bullet that came from the gun in the hand of the suspect before you. Case closed. Without recovery of live spirochetes that can divide and multiply, an argument that an infection is still active can still be made on circumstantial grounds, but it would not likely convince a jury.

A more plausible explanation for some cases of PTLDS than persisting living bacteria is that the spirochetes may be dead but their cellular remnants continue to stimulate innate and adaptive immunity for varying lengths of time. There are more convincing data for this claim.[26] Although the studies of experimental infections seldom found recoverable bacteria after treatment, it was a different story for pieces of DNA and proteins of the spirochetes. These seem to hang around in tissues long after spirochete viability is over, as detected by PCR and antibodies to *B. burgdorferi* cell parts.[27] Spirochete DNA was even taken up by ticks from the skin of infected mice and of people with Lyme disease after antibiotics.[28] In some cases, it was passed on from the ticks to other mice, but this string of events could only be detected by PCR and not by recovery of any live organisms

It is not known whether a remnant DNA is significant in terms of disease chronicity, but the conceivable importance of residual proteins for continued provocation of inflammation is more apparent.[29] The combination of residual microbe proteins and a misdirected immune response may also result in autoimmunity—the reaction of the host against its own tissues. In late Lyme disease, this is most evident among patients with arthritis of the knee or other large joint who continue to have joint swelling and pain after one, two, even three antibiotic courses. At an earlier time, the PCR test of the joint fluid or tissue may be positive. A repeat

PCR test becomes negative after antibiotic therapy, but the patient continues to have joint disease, which may take several more months to resolve. This is an uncommon occurrence now, but when it does occur, it occurs more frequently in individuals who have a distinctive genetic marker, which is also associated with a higher risk of the autoimmune disease rheumatoid arthritis.[30]

Finally, coverage of this subject would not be complete without considering two other explanations for persistent symptoms after treatment for Lyme disease: (1) coinfection with another pathogen, such as a *Babesia* parasite, carried by *Ixodes* ticks (chapter 11), or (2) coincidence with an unassociated condition in the background. In large national health surveys, about 5 percent of adults on average report fatigue as a persistent or recurrent state. That's about the same frequency as those who might have post-treatment Lyme disease syndrome.

Brucellosis: The "Chronic Lyme" of the Mid-Twentieth Century

The problem of human brucellosis in the United States is enveloped in a wide divergence of opinion and facts concerning the entity of chronic brucellosis. There are those enthusiasts who maintain that chronic brucellosis is a frequent cause of human disability, while on the other hand, there are those skeptics who do not accept the criteria that are used for the diagnosis of chronic brucellosis.[31]

Wesley Spink, 1951

Unless the diagnostician is wary he may attribute to brucellosis all the symptoms which the patient exhibits, only to find later that, in addition to Brucella *infection there is also a perhaps unrelated psychogenic or somatic illness.[32]*

Harold Harris, 1946

If the patient is presented with a diagnosis of chronic brucellosis, he will be temporarily grateful, but attempts at the treatment with

antibiotics, usually without improvement, often will plunge him
eventually into further despair and frustration.[33]

Wesley Spink, 1956

Brucellosis is a serious infection that people inadvertently acquire from unpasteurized milk products or having direct contact with infected cows, goats, sheep, or pigs. Dairy workers, ranchers, veterinarians, and slaughter-house workers are at higher risk for brucellosis. Like Lyme disease, it is a zoonosis, an infection humans get from animals, not from another person. A diagnosis of either brucellosis or Lyme disease carries no social stigma, as there might be for syphilis. As with Lyme disease, there are substantial differences in risk of the disease between geographic regions.

The infection is caused by the *Brucella* group of bacteria, which can be detected in the blood and certain tissues of acutely ill patients. As the infection evolves into a more chronic, persistent form, it becomes harder to diagnose by the direct isolation of organisms. In this case, the history, the physical exam, and a laboratory test for antibodies to *Brucella* become important to the diagnosis. (Does this sound familiar?) The antibody test for *Brucella* was probably neither better nor worse than the whole-cell-based EIA for antibodies to *B. burgdorferi*. In both tests, cross-reactive antibodies arising from other infections can result in false-positive results.

When an astute physician combined a convincing story of risk factors, compatible symptoms and physical signs, and a finding in the blood of high levels of *Brucella* antibodies, the diagnosis of chronic brucellosis was usually accurate. The definition of chronic brucellosis, however, was broadened by some physicians in the mid-twentieth century to include illnesses that might be just as easily diagnosed as "chronic Lyme disease" now or as "neurasthenia" in an earlier time. Fatigue, chilliness, mental lethargy, and muscle and joint aches were sufficient to elicit suspicion of chronic brucellosis. Blood was drawn and sent off for the *Brucella* antibody test. The criteria for a diagnosis were relaxed to include lower levels of antibodies. Such laboratory results were less reliably predictive of true infection, especially if the risk of infection was low to begin with.

It reached a point that in a scholarly book on brucellosis from 1941, the author, Harold Harris, estimated that there were 12 million cases of chronic brucellosis in the United States, when total population was around 151 million. If the infection was that widespread, it would have involved 8 percent of the population.[34] In 1941, the popular *Time* magazine's article highlighting this book said that "symptoms range all the way from mild backaches to bone and nerve infections, heart disease, and insanity."[35] For a U.S. population now at 317 million, that would translate to 25 million people afflicted with chronic brucellosis. That strains credulity, no matter how loose the diagnosis. But is an imaginative guess like that about chronic brucellosis so different from claims one can find now on the Internet of several millions undiagnosed Lyme disease cases in the United States?

With the wider application of pasteurization of milk and improved disease control measures, with quarantines of herds and vaccines for livestock, in North America, the incidence of brucellosis has declined substantially over the last several decades. As exposure to the pathogen during youth or adulthood went down, lower proportions of people were seropositive for antibodies to *Brucella*. The inappropriate use of the term "chronic brucellosis" for other disorders was eventually exposed, and the diagnosis fell out of favor—but not before many patients were given and accepted the diagnosis. Many of these people were treated with antibiotics, some for long periods, but often, as the third epigraph above suggests, in futility.

What did these people actually have? Dr. Wesley Spink and his colleagues at the University of Minnesota wondered about this, too, and considered these possible explanations: persistent low-grade infection, the remnants of the bacteria stimulating the immune system, and latent psychological problems that were laid bare by the infection. We could substitute PTLDS, and these explanations could apply as well. The phenomenon now attracting attention and controversy to Lyme disease is not something new. Odds are that there will be another attention grabber in the future as other infectious diseases with postinfection syndromes emerge and take center stage.

Chapter 11

Deer Ticks Transmit Other Diseases

In the book on Lyme disease that I published in 1996, a few paragraphs summed up the other infections that could be transmitted by *Ixodes* ticks.[1] Two decades later, this entire chapter is devoted to the subject. The other infections merit greater space now for these reasons: First, a new human pathogen, *Borrelia miyamotoi*, was discovered in the interim. Second, a type of viral infection that in the past had been associated mainly with the Eurasian continent is emerging now in North America. Third, deer tick infections like Lyme disease are spreading outward from areas where they had long been limited to. And fourth, the consequences of coinfections of *Borrelia burgdorferi* with one of these other pathogens are better appreciated.

The four deer tick–transmitted diseases considered here—human granulocytic anaplasmosis, babesiosis, tick-borne virus encephalitis, and *B. miyamotoi* relapsing fever—can be found in the same regions of North America where *B. burgdorferi* occurs. But they are not as yet as far reaching within these regions as *B. burgdorferi*. In coastal Connecticut—a center point for Lyme disease in the 1970s—*B. burgdorferi* still tops the list when it comes to prevalence in tick vectors. In this area, around 25 percent of the *I. scapularis* nymphs are infected, but the other agents are now present in ticks at frequencies of 2 to 20 percent.[2]

The chapter ends with a poorly understood, awkwardly named disease—southern tick-associated rash illness (STARI)—that is transmitted by a different type of tick. STARI mainly occurs in areas outside where *B. burgdorferi* is found, but the featured skin lesion of this aptly named "rash illness" looks remarkably enough like erythema migrans to fool many experts.

Human Granulocytic Anaplasmosis

■ **Case M.** A 78-year-old resident of northern Wisconsin was admitted to the hospital for five days because of fever, muscle aches, headache, and diarrhea. The patient had not noted any tick bites recently. His temperature on admission was 104.9°F (40.5°C). Laboratory studies of a blood sample showed that the numbers of white blood cells and platelets were lower than normal. Results of other blood tests indicated mild damage to the liver. An examination by microscope of the blood, which had been smeared on a glass slide and then stained, revealed that 10 percent of a common type of white cell, the granulocyte, had clusters of small bacteria inside them. By the polymerase chain reaction (PCR) assay, these were found to be *Anaplasma phagocytophilum*. The patient was given the antibiotic doxycycline, and his temperature returned to normal within one day. A subsequent study of his serum showed that within four weeks of disease onset, he had produced antibodies to *Anaplasma phagocytophilum* in response to the infection. At a follow-up examination three weeks after he was discharged from the hospital, he had no fever and his blood values had returned to normal.[3]

Human granulocytic anaplasmosis (HGA), like Lyme disease, is a zoonotic infection in North America that is increasing in incidence. About one thousand cases each year are officially reported in the United States—one-thirtieth of the number of official reports of Lyme disease. The accurate number of cases of HGA, if the Lyme disease example applies, may be ten times that number, or ten thousand cases a year. It was called "human granulocytic ehrlichiosis" for a time until the agent's genus name was changed from *Ehrlichia* to *Anaplasma* after bacteria family relationships became clearer. The disease is caused by a bacterium that can only live inside cells, granulocytes in particular, unlike *B. burgdorferi*, which prefers to live outside the host's cells. *Anaplasma* bacteria cannot be cultured in a medium unless the medium includes animal cells.

Granulocytes are white cells that are distinguished from lymphocytes and some other white cells by the granular appearance of the cell's insides, which gives them their name. The majority of granulocytes are "neutrophils," the white cells that make up the bulk of cells in the pus of an infected wound. Neutrophils and other granulocytes are phagocytes, whose function is to gobble up (from the Greek word for "eat," *phagein*) microbial invaders or foreign materials, such as bits of dirt in a wound or smoke particles in the lung. We can read the species name for the HGA agent, *phagocytophilum*, as a clue to the type of cell that is infected: an *Anaplasma* species that "loves" (from the Greek word for "loving," *philos*) phagocytes. Indeed, it depends on them.

The animal reservoirs for *A. phagocytophilum* are small or medium-sized mammals and sometimes birds. In the northeastern and north-central United States, common hosts are white-footed mice (*Peromyscus leucopus*), chipmunks, and shrews. The agent is transmitted by the ticks *I. scapularis* in parts of its range, *I. pacificus* in far western states, and *I. ricinus* and *I. persulcatus* on the Eurasian continent. There is no transmission of the infection from the female adult to its offspring, so nymphal ticks acquire the infection as larvae with their first blood meals. In areas where both *B. burgdorferi* and *A. phagocytophilum* are found, the latter pathogen is, on average, one-fifth as common as the former in nymphs. Unlike *B. burgdorferi*, which resides in the intestine when the tick's blood meal starts, *A. phagocytophilum* is already in the salivary glands when the tick first begins to feed.[4] It takes a while for the number of *A. phagocytophilum* to increase in those glands and prepare for the jump to a prospective host, but the rule of thumb for Lyme disease—approximately 36 hours of attachment before *B. burgdorferi* transmission can occur to a host—probably does not hold for *A. phagocytophilum*.

HGA may be asymptomatic, a mild to moderate flulike illness, or, uncommonly, a severe disease with organ failure and death (in 1 out of 100 to 200 cases). The majority of infected individuals have either no apparent disease or a mild illness. Symptomatic HGA is infrequently diagnosed in children, in contrast with Lyme disease. The incidence is highest in adults older than 40 years. The period between a known tick bite or exposure and onset of symptoms ranges between 5 and 21 days. The symptoms of chills with fever, headache, malaise, and muscle aches are not specific and may be confused with early Lyme disease when a rash is absent. The body temperature may be high, as Case M shows. Sometimes there are gastrointestinal symptoms: loss of appetite, nausea, vomiting, or diarrhea. A few HGA patients have a skin rash, but the rash is more spread out and spotty than the red patch of erythema migrans. Upwards of half the symptomatic patients with HGA infection are sick enough to be hospitalized, an uncommon event for someone with Lyme disease alone. A few patients may end up in an intensive care unit with dysfunction of multiple body systems; people of an advanced age or with an underlying compromise of the immune system are at greater risk of this outcome. A chronic, persistent infection from *A. phagocytophilum* in humans has not been observed.

A laboratory tipoff to the diagnosis of HGA is a lower-than-normal white cell count. For most bacterial infections, the white cell count is expected to be elevated. In HGA, the low white cell count is commonly accompanied by low platelet counts. Platelets are small specialized cell fragments in the blood whose job is to form part of the clot that stops bleeding from a cut or other injury. The platelet count is often abnormal but usually not so low as to put a patient at risk of severe bleeding. Blood tests may also reveal damage to the liver.

More specific laboratory confirmation is mainly through direct detection of the *Anaplasma* organisms, including a microscopic examination of blood smears stained to reveal the presence of the bacteria in a cluster (called a "morula," from the Latin for "mulberry") inside granulocytes, the polymerase chain reaction (PCR) test, or, uncommonly, cultivation in the laboratory. The PCR test is more sensitive than the microscopic exam but is available in only a few medical centers, state health department laboratories, and large commercial laboratories. The greater opportunities for direct detection of the pathogen for HGA is a major difference from Lyme disease laboratory tests, which are largely based on measurement of antibodies in the patient's blood.

Tests for antibodies to *A. phagocytophilum* exist, and these tests have their place, especially if the patient has already been started on an antibiotic for a fever of unknown cause (and as a consequence of the antibiotic, the number of the bacteria in the blood have been reduced below detectable levels). Antibodies to *A. phagocytophilum* can take time to build up in the blood, and therefore this test may not be particularly helpful at the time the patient has fever and other nonspecific symptoms. The most commonly performed antibody test is the indirect immunofluorescence assay (IFA), which requires use of a special microscope and a skilled technician for accurate performance. Moreover, the antibody test is not 100 percent specific for *A. phagocytophilum* infection; there may be cross-reactivity from infections with related bacteria that are transmitted by other types of ticks.

If a symptomatic patient has been exposed to *I. scapularis* or *I. pacificus* ticks within the prior three weeks and has an unexplained fever in combination with low white cell and low platelet counts, HGA ranks high on the possible diagnoses list. In this situation, treatment is indicated, even without laboratory confirmation in hand.[5] The antibiotic of choice is doxycycline, or other tetracycline if that one is not available.* The usual treatment duration is 10 to 14 days. This is the same dosage and duration of doxycycline as for Lyme disease, so the therapy would also be appropriate if there was an undetected *B. burgdorferi* coinfection (discussed later in this chapter). If the patient has HGA alone or with *B. burgdorferi*, the fever should not persist for longer than two days after start of doxycycline therapy. If it does, then the diagnosis is either incorrect or the patient has babesiosis in addition to or instead of HGA.

For a case of Lyme disease, there are alternatives to doxycycline for women who are pregnant or nursing and for young children: certain penicillin, cephalosporin, or macrolide antibiotics (chapter 9). For this group of individuals, a tetracycline such as doxycycline carries a potential risk of staining the teeth of unborn offspring and children without all their permanent teeth. However, a penicillin like amoxicillin or a cephalosporin like ceftriaxone are not options for treating HGA, because these antibiotics are not effective in entering human cells, where the *Anaplasma* bacteria are growing (chapter 8). Macrolide antibiotics do gain access to

---

*There was a nationwide shortage of doxycycline in the United States in 2013.

cell interiors, but they are not an option for treating HGA either, because *A. phagocytophilum* is not susceptible to them.

What to do then when HGA is suspected in a pregnant or nursing woman or in a child under age 9? Given the serious consequences of untreated HGA for some patients, and how fast the illness may worsen, there is a greater potential downside to either withholding treatment or using an unproven antibiotic than to doxycycline treatment with the possibility of stained teeth. Under these circumstances, proceeding with doxycycline therapy may be justified, according to current recommendations of the American Academy of Pediatrics and the CDC.[6] To minimize adverse effects, the tetracycline is given for as short a time as possible, usually 4 to 7 days, and is followed by a penicillin or a cephalosporin for another few days as a guard against a coinfecting *B. burgdorferi*.

Babesiosis

■ **Case N.** A 66-year-old resident of Nantucket Island, Massachusetts, was admitted to the hospital in early September after one week of fever, loss of appetite, chilliness, and fatigue. She did not recall a recent tick bite. Her temperature was 102.0°F (38.9°C), and blood tests showed she had a mild anemia. The patient's condition gradually improved, and she was discharged after five days with a diagnosis of "viral infection." However, she remained weak and easily fatigued for the next eight weeks. A sample of her blood that was smeared on a glass slide at the time of admission was reexamined with a microscope at the CDC, and *Babesia* organisms were seen inside the red blood cells. The antibody test for *Babesia* was positive. Subsequent blood smears obtained in followup were negative for *Babesia*. The patient did not receive therapy.[7]

■ **Case O.** A 68-year-old man living in eastern Pennsylvania was hospitalized in mid-August because of six days of fever, joint pains, general weakness, and confusion. He was treated for presumptive Lyme disease with antibiotics but showed no improvement. The patient did not recall a tick bite but was a gardener and participated in outdoor activities. He had never received a blood transfusion. When he was admitted to the hospital he had anemia, and the numbers of platelets in his blood were low. A smear of his blood on a glass slide was examined under the microscope, and this revealed *Babesia* parasites inside 10 percent of his red blood cells. He was treated for 5 days with clindamycin and quinine, but the fever persisted, so the therapy was changed to atovaquone and azithromycin. After 10 days of treatment, he was symptom-free, and the number of platelets in the blood had increased.[8]

Babesiosis is a parasitic disease that is similar in some features to malaria. *Babesia* organisms are protozoa, a type of one-celled animal, that infect the red blood cells of animals. While malaria parasites, which also infect red blood cells, are transmitted between vertebrate hosts by mosquitoes, only ticks transmit *Babesia* organisms. There are many species of *Babesia*, and the different species are able to infect different vertebrates. One that infects cows (but not humans) has been a major problem for the cattle industry in North America and other regions. Babesiosis of cattle has a place in the history of medicine as the first infectious disease proved to be transmitted by a blood-feeding arthropod.[9] This nineteenth-century breakthrough led to a better understanding of how malaria was transmitted between humans.

Babesia infections of humans in North America were first noted in coastal New England about the same time that a new disorder called "Lyme arthritis" was being recognized. Although the cause of Lyme disease remained a mystery for a few more years, the agent of babesiosis was there to be seen in the blood with a microscope, as Case N, which dates from the 1970s, demonstrated. The species infecting humans in the northeastern United States was identified as *Babesia microti*. The species designation derives from the genus name *Microtus* for voles, which are hamster-like rodents. While *B. microti* occurs in the two major geographic areas for *B. burgdorferi*—the northeastern and north-central United States—it has not been found in the third geographic area for Lyme disease, the far western United States. Rare cases of babesiosis occur in the western United States, but they are caused by a different species, *Babesia duncani*. In Europe the occasional cases of babesiosis are attributed mainly to another species, *Babesia divergens*.

As its species name suggests, the main reservoirs for *B. microti* are rodents, notably the white-footed mouse *P. leucopus* and chipmunks, as well as voles. Other reservoirs are shrews, raccoons, and occasionally birds. The white-footed mice commonly have the parasite in the blood, particularly during spring and summer. There is little or no transmission of the parasite from female adults through their eggs, so larvae acquire *B. microti* from infected vertebrates. There is a delay in transmission after feeding begins, because the parasite has to multiply first in the tick before it is infectious. In regions where the frequency of *B. burgdorferi* in *I. scapularis* nymphs is 25 percent or so, about 8 percent of the ticks have *B. microti*. The lower carriage rate for *B. microti* may be explained by poorer survival of *B. microti* when the larva molts to a nymph and when nymphs go through a winter season.[10]

The CDC currently records about one thousand reports of babesiosis each year, a frequency similar to HGA. With an estimate of a ten to one ratio of unreported to reported cases, the incidence can be adjusted to ten thousand cases. The frequency is increasing; there were as many cases in the last decade, 2000 to 2009, as there were in the two preceding decades combined.[11] In 2012 seven of the cases were acquired by blood transfusion, not from a tick bite. (There were no instances of transfusion-associated Lyme disease, a much more common infection.)

B. microti seems a step behind *B. burgdorferi* in its reemergence and spread from its refuges in coastal southern New England and the chain of islands from Martha's Vineyard to Long Island. The first extensive studies of babesiosis were on Nantucket Island (Case N) and Block Island, where about 10 percent of its residents had antibodies to *B. microti* in a ten-year survey completed in 2003.[12] On Nantucket and in southern Connecticut, the ratio of Lyme disease cases to babesiosis cases is about three to one, while in northern Connecticut, up the Connecticut River valley, the ratio is closer to thirty to one.[13]

The infection with *B. microti* ranges in severity from asymptomatic to fatal.[14] Most infected persons who come to medical attention have a febrile illness with many of the same symptoms as with early Lyme disease or HGA: chills, sweats, muscle aches, joint pains, and fatigue. These symptoms of babesiosis are similar to the symptoms of a viral illness that may be going around. What increases the suspicion of babesiosis is the physician's finding on the physical exam of an enlarged spleen or liver to go along with the fever. Routine laboratory tests may reveal an anemia and low numbers of platelets.

Children more commonly than adults have either an asymptomatic infection or an illness so mild as to merit little concern. Most symptomatic cases are in people over the age of 40, and the highest incidence of the illness is in those over 65. About half of people who have babesiosis are hospitalized. Those who are over the age of 50 or who have a deficiency in their immunity are more likely to have a severe course and a fatal outcome from babesiosis. One particularly disadvantageous condition for *Babesia* infection outcome is absence of a spleen, which may have been surgically removed following abdominal trauma or for other reasons. The spleen is an organ that effectively filters infections in the blood and can remove red cells that have the parasites. When a person lacks the ability to clear these affected red cells, the numbers of parasites in the blood can mount to dangerous levels and overwhelm the other defenses.

One feature that differentiates babesiosis from Lyme disease and HGA is persistent circulation of the parasite in the blood in some individuals after the acute febrile illness has resolved. This can occur even in people who either were asymptomatic or did not come to medical attention. This is the explanation for cases of babesiosis where the only risk factor was receiving a blood transfusion.[15] Over one hundred cases of babesiosis transmitted by transfusion of blood or its products have been documented. *B. microti* can survive and remain infectious inside blood that is stored at cold temperatures for weeks at a time in blood banks. The fatality rate for transfusion-associated cases is higher than for those acquired by tick bite, perhaps because people who get blood transfusions for reasons other than trauma often have underlying illnesses.

This is the bad news about babesiosis. The good news is that a blood test can directly detect *Babesia*, and this test can be carried out in most hospital laboratories. In other words, the specimen need not be sent off to a regional or national laboratory at a distance for specialized testing. The parasite's dependence on red cells to live, and the abundant number of parasites in the blood, mean that in most cases of symptomatic babe-siosis, the telltale signs of the protozoa inside the red cells can be seen by conscientious microscopic examination. The technician or lab manager patiently looks over glass slides with blood smears that have been spe-cially stained to enhance the parasite's appearance. But first the diagno-sis needs to be suspected. Blood smear exams for blood parasites are not routinely requested for febrile illnesses in the United States or Canada, unless the patient has recently returned from travel in a country with ma-laria. Since the patient who has babesiosis is probably also at risk of HGA, other blood smears may be scrutinized for evidence of HGA.*

All of this assumes that the hospital or clinic has experienced and trained personnel who can accurately perform the blood smear examina-tion and can distinguish true *Babesia* cells from the occasional but in-evitable spots on the blood cells that are byproducts of staining and pro-cessing the slide. This is a good bet for hospitals where the medical and laboratory staff are well acquainted with the disease but is not necessar-ily so for hospitals outside these areas. When local expertise is not avail-able, the smear or blood sample can be sent to a reference laboratory that offers the test.

These regional and national laboratories would also likely offer a PCR test for *B. microti*, which can be as sensitive as a careful and expert blood smear exam. A PCR-based assay for this disease, as for Lyme dis-ease and HGA, has not been approved by the FDA and may not merit health insurance company reimbursement. The laboratory that performs the PCR should not only demonstrate experience with this test but also meet the highest professional standards, including certification by in-dependent quality control assessments.

As was true for Lyme disease diagnosis, there are tests for antibodies to *B. microti*, but at present this is the more complex IFA test rather than an EIA. Moreover, the outcome is subject to the time it takes to mount an immune response to the infection. Since most antibiotics do not affect *Babesia* parasites, there is less chance that previous treatment of a

---

*Microscopic detection in blood smears is not feasible for *B. burgdorferi* infection, even when spirochetes are likely present in the blood, because there are so few of them.

febrile patient with an antibiotic will invalidate the direct detection assay, be it a blood smear or a PCR test. And so there may be less need to fall back on an antibody test for confirmation of babesiosis than is true with HGA. In any case, the presence of antibodies to *B. microti* does not mean that there is an active infection; it only indicates a likelihood that the patient has had *B. microti* infection sometime in the past if not the present. Antibodies persist long after an infection has resolved. Since *B. microti* infection is more likely to be asymptomatic or undiagnosed, especially in children and young adults, than is *B. burgdorferi* infection in North America, many people who grew up or lived in high-risk areas have detectable antibodies to the parasite. This does not imply an active infection; it does not mean that babesiosis explains any current symptoms. (Whether persons who have ever had babesiosis should be blood donors is another matter and is the subject of study and deliberation.)

Treatment of active babesiosis—defined as applicable symptoms plus documented presence of parasites in the blood—calls for the use of two drugs at the same time. Combination therapy is common practice for treatment of serious parasitic diseases of the blood, because one drug by itself may not be sufficient. Babesiosis therapy has more in common with therapy for malaria than with therapy for Lyme disease or HGA. Except for the inclusion of a macrolide in one of the two recommended treatment regimens, the medications are completely different and would not be expected to have any effect against *B. burgdorferi*. Both *Babesia* treatment regimens include an antimalarial drug: atovaquone with the macrolide azithromycin is one treatment, and quinine with another type of antibiotic, clindamycin, is the other. Quinine, the original therapy for malaria, can have serious side effects in the doses used, so the regimen of quinine and clindamycin is generally reserved for severe cases or if atovaquone and azithromycin have failed. As Case O shows, sometimes it is the quinine and clindamycin that fails, and the alternate regimen succeeds. The duration of therapy for both regimens is 7 to 10 days. Clinical improvement typically occurs before the parasites are completely eliminated from the blood.

Treatment for babesiosis by either regimen carries more risks than oral treatment for Lyme disease and should not be started until the patient has been informed of what the side effects may be. Anti-*Babesia* therapy is usually not recommended for people who may have antibodies to *B. microti* but have no parasites in the blood by microscopy or PCR. On the other hand, treatment may be indicated for persons without symptoms or signs of disease but who have had the parasites in their blood for more than three months.

Deer Tick Virus Encephalitis

■ **Case P.** A 62-year-old resident of New York was admitted to a hospital in late spring after four days of fever and fatigue followed by double vision, difficulty speaking, and weakness of the right arm and leg. He owned horses and spent time in the outdoors in an area where Lyme disease was common. He had been under treatment for the last four years for a chronic form of leukemia, but had otherwise been healthy. Tests for antibodies to *B. burgdorferi* and *A. phagocytophilum* were negative. After being admitted to the hospital, he was treated with a variety of drugs against bacteria and viruses for suspected encephalitis. But he continued to have a high fever (104.5°F, 40.3°C) and a progression of the disease of his nervous system. He died seventeen days after his symptoms started. An autopsy revealed widespread infection of his brain by the deer tick virus and severe inflammation.[16]

Case P's illness was devastating and ultimately fatal. He had encephalitis, or brain (encephal-) inflammation (-itis). The specific diagnosis came after death, but it is doubtful that much could have been done to prevent death even if the diagnosis had been made earlier. There are few options for therapy other than trying to keep the patient alive and comfortable while hoping that the immune system overcomes the virus before too much damage has been done. Deer tick virus encephalitis is still rare;* between 2001 and 2012, there were, by the CDC's accounting,

---

*Earlier cases of this encephalitis may have been incorrectly labeled as "Powassan virus encephalitis."[17] Powassan virus is closely related to the deer tick virus but is usually transmitted by ticks that uncommonly bite humans.

forty-five cases, most of which were from Minnesota, Wisconsin, and New York. But the trend is an increasing number of cases in the northeastern and north-central United States. Some of this trend may be attributable to better awareness of the disease by medical practitioners. But in what may be a harbinger for human risk, there has been a substantial rise in frequency of infections in deer over the last three decades in New England states.[18]

The deer tick virus is a small virus that is related to agents that cause West Nile encephalitis, dengue fever, and yellow fever. The virus is already in the salivary glands of the unfed tick and may therefore be transmitted early during feeding. From 2 to 4 percent of *I. scapularis* ticks in Connecticut were found to be infected in a recent survey.[19] A major vertebrate reservoir is the white-footed mouse *P. leucopus*,[20] but other mammals may serve this role as well. In humans and other animals that are susceptible to severe disease, the virus first invades cells in the skin, then spreads to the lymph nodes, where it replicates, and from there makes it way to the brain. The inflammation and damage in the brain may be so great as to be fatal, as in Case P, or may lead to permanent neurologic deficits if the patient survives.

In Europe and Russia, there is another viral encephalitis, called tick-borne encephalitis (TBE), that has caused hundreds of thousands of cases of human disease on that continent over several decades. The TBE virus is related to the deer tick virus, is transmitted by *I. ricinus* and *I. persulcatus* ticks, and uses rodents as reservoirs. The larger experience in Eurasia with the TBE pathogen offers lessons for what North America might expect here with the deer tick virus.

Currently there are annually about twelve thousand TBE cases, most of which are in Russia and Eastern and Central Europe. There is a higher incidence in males and those aged 17 to 40. Agricultural and forestry workers as well as those with recreational exposures to forested areas have a greater risk of infection. In regions where the TBE virus is established in the environment, the odds of getting encephalitis from a single tick bite are about one in one hundred.

For the majority of individuals who get infected, there are either no symptoms or symptoms are too mild to be easily recalled. For those who become ill, the onset of symptoms is one to two weeks after exposure to *Ixodes* ticks. The usual initial form of the illness is sudden onset of fever, muscle aches, nausea, and severe headache that lasts for up to a week. There may be neck stiffness and pain, an indication of involvement of the covering of the brain. Most patients recover from this point, while others continue to have fever, nausea, and headache. A third, smaller group goes on to show effects on the central nervous system, with signs of encephalitis: delirium with hallucinations, drowsiness that might progress to coma, or stroke-like effects, with weakness on one side of the body. The mortality rate is 4 to 25 percent. Among the survivors, recovery can be slow, with a fatigue syndrome and sometimes changes in mood and personality. One in five patients has residual effects, including paralysis, incoordination, or difficulty thinking and concentrating. Some patients seem to have recovered from the fever and other symptoms, but go on to develop a polio-like disorder, with partial paralysis. The affected muscles may atrophy, and half of these patients have residual muscle weakness.

There are no findings from routine laboratory tests that would obviously point to the virus. The number of white cells in the blood may be low, but that can occur with HGA or babesiosis as well as with several other viral infections. If there are signs of invasion of the brain or its covering, a lumbar puncture is usually performed. The results may reveal increased numbers of white cells in the cerebrospinal fluid, and the virus may be detected by culture or by PCR. Tests for antibodies to the TBE virus may be positive when the patient first sees the physician, because the infection may have been going on for two to four weeks before neurologic symptoms and signs appear—enough time for the antibodies to develop and be detected.

There is at present no antiviral drug—the equivalent of an antibiotic but for viruses—for TBE. In Russia and Europe, some patients have been treated with sera that has been obtained either from individuals who have recovered from the infection or from horses that have been immunized with the virus. But the serum treatment has to be administered within the first two days of disease to have much effect. What is available—and has been for many years—is an effective vaccine against TBE virus infection in Europe and Russia. One of the formulations, which is widely used in Austria, gives a protection rate of over 95 percent after the three recommended doses. The protective effect of the vaccine seems to last for at least three years.[21] Whether the TBE vaccine would also provide protection against the deer tick virus is not known, but it is unlikely, given the differences between the two viruses in their proteins.

Relapsing Fever

■ **Case Q.** A 61-year-old resident of Massachusetts, who previously was in good health, was admitted to a hospital in August after two days of feverishness and shaking chills. He also had headaches, visual sensitivity to bright light, muscle aches, joint pains without swelling, and loss of appetite. During the intake, he stated that he had been playing golf near the coast but had not noted any tick bites. In the hospital, the patient continued to have shaking chills followed by drenching sweats and temperatures up to 102.9°F (39.4°C). There was no skin rash or other abnormalities on physical exam, but the laboratory tests of the blood showed low numbers of white cells and platelets and evidence of liver damage. Microscopic examination of blood smears did not reveal *B. microti*. He was treated with the antibiotic doxycycline by vein for presumptive HGA, and his temperatures returned to normal over the next three days. He continued to take this antibiotic in oral form after discharge for a total of two weeks. On followup, he was asymptomatic, and his blood test values had returned to normal values. Subsequent study of blood taken on admission revealed by PCR the presence of the relapsing fever agent *B. miyamotoi* in the blood but not *B. burgdorferi*, *A. phagocytophilum*, or *B. microti*.[22]

In 1868 Berlin Otto Obermeier, a 25-year-old physician, was looking at a patient's blood with a microscope and noted "minute organisms presenting a twisting or rotatory movement."[23] This was the first known observation of a *Borrelia* spirochete. The blood came from a patient with relapsing fever, an epidemic disease in Europe at the time and characterized by recurring bouts of fever. The organisms Obermeier described in his subsequent article were eventually named *Borrelia recurrentis*. Over the next decades, this species was joined in the genus *Borrelia* by several others, such as *B. hermsii* in North America, which causes a tick-borne version of relapsing fever. At the peaks of body temperature, numerous spirochetes are swimming among the red cells of the blood of patients.[24] Eventually the fever relapses stop, and the patient recovers, usually without any complication or long-term sequelae.

The ticks that transmit most types of relapsing fever are in a different category than deer ticks and similar ticks, like the dog tick, that humans regularly encounter. These more typical vectors of relapsing fever are known as "soft ticks" to distinguish them from the more sculpted, less squishy "hard ticks" typified by *I. scapularis*. The relapsing fever ticks differ from these other ticks in their behavior as well as their body style: most kinds of soft ticks are seldom seen by humans, because they live in close proximity to their intended hosts and usually come out to feed only at night while people are sleeping. There was one relapsing fever *Borrelia* species known to be transmitted by a hard tick, but for a long time that particular tick was thought to be the sole exception.

A second exception was identified in 1995 when *B. miyamotoi* spirochetes were discovered in the hard tick *I. persulcatus* in Japan.[25] This was after the Lyme disease species had been identified, so the new organism could be compared with those *Borrelia* species as well as with the relapsing fever group. It turned out that *B. miyamotoi* is solidly in the relapsing fever group of species.[26] Eventually *B. miyamotoi* was found in *I. scapularis* ticks in the northeastern and north-central United States, in *I. pacificus* ticks in the far western United States, and in *I. ricinus* ticks in Europe. Where *B. miyamotoi* occurs in nature, its average prevalence in nymphs is about 2 percent, one-tenth that of *B. burgdorferi*. The white-footed mouse is commonly infected and is a major reservoir.[27] Unlike *B. burgdorferi*, which transiently circulates in the blood in low numbers, *B. miyamotoi* cells reach much higher densities in the blood of animals in nature and in the laboratory. But it did not look like the organism was a cause of human disease.

That might have been the end of the story on *B. miyamotoi*: a biological curiosity of little relevance for medical care. But it wasn't. First in Russia and then in North America and Europe, cases of human illness caused by *B. miyamotoi* started being reported. The illnesses range from a febrile illness that could be confused with HGA, as in Case Q, to more serious conditions with involvement of the brain and its covering. A survey of serum samples collected from individuals living in southern New England revealed that about 4 percent had antibodies to *B. miyamotoi*, compared with 9 percent who had antibodies to *B. burgdorferi*.[28] This suggests that *B. miyamotoi* infection of humans is much more common than has been appreciated, but also that the infection is mild enough not to stand out for many affected individuals or their health care providers.

For laboratory diagnosis there is the option of looking for the spirochetes in a blood smear under the microscope, as is commonly done for other forms of relapsing fever. An antibody test has been developed for *B. miyamotoi*, and it is becoming available for use beyond the research lab. The test is based on a purified protein found in relapsing fever *Borrelia* species but not in Lyme disease species, so it could be used to discriminate between the two infections. One of the challenges that the recognition of *B. miyamotoi* relapsing fever poses is the possibility of antibody cross-reactivity between *B. miyamotoi* and *B. burgdorferi*. At one time cross-reactivity between relapsing fever species, such as *B. hermsii*, and Lyme disease species in antibody tests, while experimentally proven, was not thought to be of practical concern because the geographic areas for the two diseases in North America and Europe minimally overlapped. The finding that both infectious agents have the same tick vector and that *B. miyamotoi* occurs wherever *B. burgdorferi* occurs, albeit at a lower prevalence, means that the specificity of the Lyme disease antibody tests has to be reevaluated. This recognition may spur the development of more *B. burgdorferi* tests that are based on purified proteins, not whole cells, and that will be more specific for Lyme disease species.

The final word for this section is on treatment, and until more studies are done, it is short. The recommended antibiotics, such as doxycycline and amoxicillin, for Lyme disease should also be efficacious for *B. miyamotoi* infection. This includes the use of an intravenous antibiotic, like ceftriaxone, when the infection has invaded the brain. These antibiotics have good track records for treating other forms of relapsing fever.

Coinfections

When the same ticks are the vectors for at least five different infections, then infections by two or three at the same time are to be expected in some cases. There may be some clustering of the microbes in a single tick to a greater extent than would be expected by random odds, but for the most part, the infectious agents are spread out among the ticks in a given environment in North America and Europe.[29] That means, for example, that if *B. burgdorferi* is in 25 percent of nymphs and *B. microti* is in 8 percent, about 2 percent of the nymphs would be expected to have both. If we want to add in *A. phagocytophilum* at 5 percent prevalence, only one out of one thousand ticks on average would have all three pathogens. *B. burgdorferi* and *B. miyamotoi* occur together in *I. scapularis* nymphs no more often than chance would predict.

In studies of humans with early Lyme disease and erythema migrans, coinfections with either *B. microti* or *A. phagocytophilum* were documented, but at frequencies that corresponded to calculations based on incidences of single pathogen infections. For instance, in one study of ninety-three patients with culture-proven *B. burgdorferi* infection, two patients (2 percent) also had active infection with *B. microti,* and another two patients had coinfection with *A. phagocytophilum.*[30] None had all three infections at the same time.

When coinfections do occur, they can lead to more severe and longer-lasting disease than is seen with single infections.[31] Patients with active *B. burgdorferi* and *B. microti* infections tend to have an enlarged spleen and more pronounced fevers and other systemic symptoms, like muscle aches, than patients with Lyme disease alone. Accurate diagnosis of coinfection of *B. burgdorferi* and *B. microti* is especially important because of the differences in therapies for each infection. Persistence of symptoms after what should have been adequate treatment for Lyme disease may be explained by a coinfecting *B. microti* parasite.

What about *Bartonella*, a pathogen sometimes invoked as an explanation for persistent symptoms after Lyme disease treatment? This is another bacterium that infects the blood. Like *Babesia, Bartonella* bacteria infect red cells. Various species of *Bartonella* are commonly found in the blood of wild and domestic animals, including cats, but they are usually transmitted between their mammalian reservoirs by fleas. Occasionally, traces of *Bartonella* bacteria have been found in *Ixodes* species ticks. However, there is not convincing evidence that *Bartonella* can be transmitted to people or to other animals by *Ixodes* ticks in North America.[32]

STARI (Southern Tick-Associated Rash Illness)

■ **Case R.** On Long Island, New York, in mid-May 2009, a woman removed an embedded tick from the back of her 2-year-old child. A week later, the child's mother noticed a half-inch (1 centimeter) diameter circle of redness at the bite site. The next day the child was seen by her pediatrician, who noted a 1-inch-wide circular skin rash on her back, but that she was otherwise well and without fever. Erythema migrans of Lyme disease was diagnosis, and the child was treated with the antibiotic amoxicillin by mouth. Over the next six days, however, the skin rash continued to expand despite the therapy. On repeat exam, the child again appeared well, except for the circular rash on the back that was now 3 inches (8 centimeters) across. Because of the unaccountable lack of effect of the amoxicillin, a skin biopsy and blood sample were obtained. Microscopic study of the skin showed inflammation but no evidence of spirochetes. A culture of the biopsy was negative for *B. burgdorferi*. Tests of blood, including white cell counts, were normal, and there was no detectable antibody to *B. burgdorferi*. The amoxicillin was continued. After the 7th day of treatment, the rash began to fade and did not return after the antibiotic was stopped after 20 days total. The family had saved the tick that bit the child, and it was identified by an entomologist as an adult Lone Star tick (*Amblyomma americanum*). A PCR test of the tick for *B. burgdorferi* was negative.[33]

Case R had an expanding circular rash at the site of a tick bite on Long Island in late spring. Sounds like early Lyme disease, right? But for one difference, it could have been. The single difference: it was a Lone Star tick, not a deer tick, that had bitten the child. Based on this association, and the lack of evidence of *B. burgdorferi*'s presence, the diagnosis of southern tick-associated rash illness (STARI) was made. If the patient was old enough to give an account, we might have heard of a mild fever, headache, muscle aches, or joint pains accompanying the rash. Although there was no discernible response to the amoxicillin after a week, eventually the rash illness subsided, and the patient suffered no aftereffects.

The essential clinical features of what came to be called STARI were first described by a family practitioner, Edward Masters, of Cape Giradeau, Missouri.[34] There is a solitary rash that looks like erythema migrans but for a few subtle differences.[35] The same early general symptoms, like chills and muscle aches, might appear as well, but not the later manifestations of Lyme disease in joints, nervous system, or heart. Tests for antibodies to *B. burgdorferi* or other known tick-borne pathogens, like *Rickettsia rickettsii*, the bacterium that causes Rocky Mountain spotted fever, are negative. The same oral antibiotics that are used to treat Lyme disease appear to shorten the course of the illness, but there have been no clinical trials.

Masters and others noted a strong association of STARI with either a bite or an exposure to *Amblyomma americanum*, called the Lone Star tick for the single white spot on the back of the adult female. The tick's common name also hints at the geographic range of the tick. Texas, the Lone Star state, occupies a considerable portion of its geographic footprint, which has mainly been in the southeastern and south-central states. But as Case R exemplifies, its range now includes the mid-Atlantic states, like New York, and southern New England. So, the "southern" part of STARI may be misleading, even more so if a trend toward northward migration continues.[36] Unlike the southern form of *I. scapularis, A. americanum* ticks are not reclusive or shy about biting people.

What causes STARI remains a mystery. It behaves like an infectious disease and seems to respond to antibiotics. But the usual signatures of a pathogen at work, such as passage of a disease from one animal to another or isolation of a microbe, are lacking. Colleagues and I discovered in Lone Star ticks a new species of *Borrelia* and named it *B. lonestari*.[37] *B. lonestari* can be found in the blood of white-tailed deer in the southern United States, and it infects birds, such as wild turkeys, as well. But so far there is not proof that it can infect humans, let alone cause

STARI. Other proposed candidates for the agent have not panned out yet either.

A. americanum is not capable of transmitting *B. burgdorferi*,* but it is a competent vector for the agents of other human diseases, one of which is human monocytic ehrlichiosis (HME), caused by *Ehrlichia chaffeensis*, an organism related to *A. phagocytophilum*. HME is similar enough to HGA to be clinically indistinguishable; the main differentiating points are the targeted white cell, the vector tick, and the geographic range for the disease. An even more deadly infection associated with Lone Star ticks is heartland virus disease, which has emerged in the last few years in rural areas where bites from this tick are common.[38]

In late August 2006 President George W. Bush was treated for what was thought to be Lyme disease. According to a 2007 *Washington Post* article that first reported this illness a year after the incident, the president had "found a rash on the front of his lower left leg and alerted White House physicians."[39] Mr. Bush had vacationed as usual the first part of August 2006 at his ranch in Crawford, Texas.[40,41] Tony Snow, the White House press secretary, said that Mr. Bush spent August 4, his second day at the ranch, "brush cutting and trail clearing, followed by a bike ride."[42] From these facts, my view is that Mr. Bush's illness was STARI and not *B. burgdorferi* infection. The Edwards Plateau of the plains of central Texas features white-tailed deer and wild turkeys, favored hosts for the Lone Star tick, and in early August the nymphal ticks would have been active. *I. scapularis* ticks can be found in that part of Texas,[43] but as discussed in chapter 3, the southern form of this species shows little inclination to bite humans. Moreover, there is no convincing evidence that *B. burgdorferi* was present in Texas in 2006. Cases of what was reported as "Lyme disease" for Texas in 1992 through 1996 (figure 4.3 in chapter 4) had dropped in number in the 2007–2011 period, in apparent recognition that these were locally acquired cases of STARI and not cases of Lyme disease.

---

*Early reports of "*B. burgdorferi*" in *A. americanum* ticks were based on studies with nonspecific reagents; the observed spirochetes were *B. lonestari* instead.

Chapter 12

Preventing Lyme Disease
Community-Wide Measures

In the Soho district of London, 39 Broadwick Street has long been the home for a pub named the John Snow. On the sidewalk outside the drinking establishment, a pink granite slab stands to commemorate where a public water pump was located 160 years ago. In August 1854, a cholera outbreak raged in the surrounding neighborhood. The first case was at 40 Broad Street, as Broadwick Street was then called. Over a three-day period around the end of August, more than a hundred people living around or on Broad Street died from the severe diarrhea and dehydration of cholera. John Snow, a 41-year-old physician who lived in the area, had a few years before published an article proposing that sewage-contaminated drinking water was the source of cholera.[1] This conjecture was dismissed by medical authorities (as well as by the water companies), who favored the then-popular explanation of a "miasma" in the air. But Dr. Snow was not deterred. He began an investigation of the outbreak occurring in his neighborhood and recorded that most of the deaths occurred within a short distance of the Broad Street pump, which was served by one of the private water companies then operating in London. There was little or no cholera among Soho residents who got their water from another water company (the control group, we would term it). With this evidence in hand, Dr. Snow convinced the local parish's officials to remove the handle from the Broad Street pump. The Soho outbreak subsided thereafter, and that pump is now famous in the annals of epidemiology. When an investigation of an epidemic locates its source, and intervention is called for, the metaphor used is "removing the pump handle."

What is the equivalent of the "pump handle" for Lyme disease? *Borrelia burgdorferi*'s life cycle is more complex than the cholera pathogen's. Humans are both the victims and the reservoirs for cholera, and transmission occurs through the fouling of drinking water. Would-be interveners against Lyme disease have both the animal reservoirs and the arthropod vector to contend with. A disease better suited for comparison with Lyme disease is malaria. Similarities and differences between Lyme disease and malaria are pointed out throughout this chapter and the next. To start, Lyme disease and malaria are alike in being transmitted by blood-feeding arthropods: ticks for Lyme disease and mosquitoes for malaria. On the other hand, while humans are the critical reservoirs for malaria parasites, we are irrelevant for maintaining *B. burgdorferi* in nature.

Another theme organizing this and the next chapter is the distinction between a community-wide—that is, a public health—intervention and an action that persons or households can do on their own. The decision to remove the Broad Street pump's handle was made by a group of people with the entire neighborhood in mind. This community action presumably was not without a temporary cost: people had to make a longer trip to get water at another source, for instance, until the water supply was attended to. In the same outbreak in 1854—an era before the germ theory was accepted—many Soho residents recognized that there was something dangerous about where they were living. Those with the means left the city, as their forebears might have done when plague had broken out in London two centuries earlier. That individual action might have saved them and their family members who traveled with them, but it would have only marginally altered the course of the outbreak for those who stayed put in Soho. Reducing the ranks of susceptibles was not as effective as Dr. Snow's community-wide prescription for ending the epidemic in Soho.

Now, back to the pump handle question, rephrased this way: If you were awarded one million dollars with the constraint that it could be used to fund one, just one, project to prevent *B. burgdorferi* infection in your community, what would it be? The success of your project will be judged by this criterion: the reduction in the incidence of Lyme disease after five years.* Let's set aside for now going after the pathogen itself. Many successful disease control measures, such as separating sewage from drinking water, were instituted before the microbes that cause diarrheal diseases were recognized. Later, we'll consider possible interventions directly against the pathogen, such as a vaccine.

Consider then three candidates for a "pump handle" for Lyme disease: one, the tick that transmits the pathogen; two, the white-footed mouse that is the pathogen's reservoir; and three, deer, which are the hosts for adult ticks. If the Broad Street pump delivered the cholera microbe to those London residents, then the *Ixodes scapularis* tick is what delivers *B. burgdorferi* to people in Lyme, Connecticut, and elsewhere. One could argue that the water company was the culprit in the 1854 outbreak, and then, by similar reasoning, implicate infected white-footed mice (or other small mammals) as the true source of the problem today. Maybe so, in a chicken-or-egg sense. For John Snow the expedient action was to remove the pump handle. What was the alternative—trying to convince a skeptical water company to change its practice? Adapting this approach to Lyme disease, we arrive at this objective: reduce bites by deer ticks, either by decreasing the number of ticks or by interfering with ticks before they can attach and transmit.

A more nuanced position is that infected nymphal ticks, rather than ticks in general, largely determine risk. A strategy based on this view entails targeting only infected ticks and somehow sparing the uninfected ones. This surgical approach makes theoretical sense, but it means going after the pathogen itself. As we will see, it is one thing to bolster the defenses of informed people against *B. burgdorferi* with a vaccine or to kill the invading microbe with an antibiotic after a tick bite; it is another matter to carry out an immunization or treatment program for wildlife on public and private lands that are not your own.

So we start with options for control of deer ticks in general. The list of measures to achieve this goal includes those that exploit other members of the life cycle: the small mammals and deer. In the absence of an

---

* A typical National Institutes of Health research grant is about $1 million over 5 years.

effective human vaccine, a sustained reduction in Lyme disease risk for residents and visitors is probably best achieved by a concerted action at a community or municipality level, the larger in scope the better.

To be sure, there are obstacles to accomplishing actions at the community or broader level. When proposed remedies include culling deer, area-wide application of insecticides, or erecting a high fence around a town, we enter social and political realms where there are differing points of view and constituencies.[2] Quarantine, mandatory vaccination, and other measures that citizens allow governmental health agencies to take in efforts to control contagious human diseases are not plausible options for preventing Lyme disease. Since *B. burgdorferi* is not spread person to person, there is not the same public health impetus for government involvement as there might be for human-transmitted infections like measles, tuberculosis, or whooping cough. A school may be closed if there are cases of measles or whooping cough. A patient who refuses to take tuberculosis medications may be confined in a hospital under a court order. In 2014 health authorities in West Africa instituted quarantines for control of the Ebola virus epidemic. But someone who has Lyme disease is not a threat to other people, at least not on contagiousness grounds. If a town has a large burden of Lyme disease among its residents and visitors, it can look to local, state, or federal public health agencies for advice and expertise, but not, as things stand now, for imposition of the types of sanctioned measures for disease control that come into play for an outbreak of food-borne hepatitis or meningitis. It's largely up to a municipality or other self-governing locality to take the initiative, if there is the political will.

Lessons from the Malaria Eradication Campaign

Toward the end of the nineteenth century, most parts of the United States had malaria.[3] This included California during the gold rush and north into New England, where Lyme disease now is the number one arthropod-borne disease. Deforestation and early industrialization began the retreat of *B. burgdorferi*, deer, and *I. scapularis* to a few refuges. During the same period, other landscape changes, such as an increase in ponds, pools, ditches, and canals, favored the growth of mosquito populations and brought malaria into areas it had never been before. As these lands were in turn converted over time to farms for crops and cows and other domestic animals, there was decline in mosquito populations and malaria incidence in many states, including in the northeastern and north-central regions. But malaria continued to be a major health problem in the southeastern states well into the twentieth century. In a 1933 survey of schoolchildren in southeastern states, about 6 percent were infected with the malaria parasite.*

Something had to be done. It was time to eliminate malaria from the United States. This effort entailed cooperation between the CDC of the U.S. Public Health Service and state and local health departments in thirteen states that were still "malarious."[4] The mosquito or vector control agencies or districts that were created during that period are still in existence as part of state and local governments. The early efforts included widespread terrestrial and aerial spraying of diesel oil and insecticides, like arsenic compounds, on ponds and wetlands to kill the mosquito larvae, as well as drainage projects and mosquito-proofing houses, such as with window screens. These measures were partially successful, but spraying of pesticides throughout the countryside was considered a blunt instrument. The leaders of the eradication effort concluded that it was more efficient to target the small proportion of mosquitoes that were liable to bite people instead of all the mosquitoes in the wider area. This meant going after those populations living around houses and other residences. If the mosquitoes that picked up the parasite from an infected person could be killed before they had an opportunity to bite a second person, then malaria could not be transmitted.

The method that worked was spraying the insecticide DDT (dichloro-diphenyl-trichloroethane) on the interior walls of about 2.5 million homes throughout the region. DDT is called a "residual" insecticide, because it continues to have effects against insects—as well as birds and

---

*This was not the deadly *Plasmodium falciparum* type of malaria but the less serious yet more chronically disabling form called *Plasmodium vivax* malaria.

mammals—weeks to months after the application (and for this reason and its overall toxicity, DDT is highly restricted now). Only a few applications to each house were necessary. By around 1950, malaria transmission within the United States was eliminated. There are still cases of malaria on U.S. soil, but these invariably have been acquired outside the continental United States during travels or overseas service.

The local and state vector control agencies that were integral to the campaign's success continue to exist in the southeastern United States as well as most other states. Their mosquito control efforts are now directed more to serious viral diseases, like West Nile encephalitis or dengue fever, transmitted by these insects.* Weapons against mosquitoes now include other options besides pesticides, such as larvae-eating fish and bacteria that make mosquitoes sick, two examples of "biological control" agents. There are also other chemicals, like [S]-methoprene, that do not directly kill the larvae but interfere with their development into adults by falsely mimicking an insect hormone.

But applications of insecticides to ponds, irrigation ditches, canals, marshlands, and elsewhere for killing larvae are sometimes still necessary. Insecticides are also used by state and local vector control agencies to go after adult mosquitoes in the air, and may be dispensed as low-dose aerosols by aircraft and sprayers on trucks. When the risk of a serious disease like viral encephalitis or dengue fever increases, these measures may be called for. To be sure, these are safer insecticides that will not linger, unlike the now-restricted DDT, in the environment or in wildlife, pets, or people that are inadvertently exposed. Yet they are pesticides nonetheless, with toxicities and various effects on nontarget species.

To what extent is the malaria experience relevant for Lyme disease control? A common feature of malaria and Lyme disease is that there is no vaccine option to protect humans against infection or to reduce the pool of infected reservoirs (humans for malaria and small mammals for Lyme disease). Another similarity is the major importance of personal protection measures for prevention. These measures include window screens and

---

*These special health agencies may also deal with rats, fleas, flies, fire ants, ticks, and other pests.

bed nets for mosquitoes, suitable clothing and body inspections for ticks, and repellents for both (chapter 13). For malaria and Lyme disease alike, there are prophylactic possibilities—taking a drug to head off infection rather than treating it after it's under way.

Beyond the similarities between these vector-borne diseases are two important differences that limit the applicability of malaria's lessons. The first is the distinction between the key reservoirs for *B. burgdorferi* and the prospective victims we are trying to protect. Cumulatively preventing malaria in one person after another in a population can result in lowering the overall risk for a community, because there would be a lower density of parasitized humans to infect the next round of mosquitoes that are born. That's one reason why malaria disappeared in parts of the United States before there were effective therapies. That outcome is probably not the future for Lyme disease. Even if every person in town was magically shielded from *B. burgdorferi* infection for one year, as long as there are infected rodents and ticks in the environment, once the spell was lifted, the risk of infection would return to what it was. The second difference is the greater leeway for mosquito control on a community-wide level than for tick control. This is partly because of the greater difficulty in finding and selectively targeting ticks than mosquitoes in the environment. Mosquito development includes a water phase, which allows for efficient dispensing of compounds to the target, and mosquitoes are vulnerable to aerial spraying while they are flying around. It also seems that the public is more accepting of the area-wide use of insecticides against mosquitoes than against ticks.

This chapter and the next chapter describe prevention options for Lyme disease. While the emphasis is on what's realistic, some less developed and yet-to-be-tested strategies are included. Some of these strategies show promise and may become preferred methods in the future. Over time, the underpinnings for each of the options will likely change little, while the specific recommendations and particulars for pesticides, repellents, host targeting devices, and so on could change. Thus, explanations and concepts are the focus in these chapters, rather than brand names and dosages. The "Trusted Internet Sites" section lists some sources for obtaining up-to-date specifics on what communities and individuals can do, particularly in ways best suited for their own locales.* I start in this chapter with community-wide measures before considering personal and household protection in the next chapter. This distinction blurs when some of the tools, such as landscape treatments, can be scaled up or down in their spatial scope, from state or national forest to residential backyard.

Keep in mind a common finding of many research studies covered here (or searchable elsewhere): an intervention's measurable outcome, like deer population size or nymphal tick density, does not necessarily translate into quantifiable progress toward the primary goal, such as reducing Lyme disease incidence. In other words, what is viewed as a significant effect of an intervention—say, a 30 percent drop in infected ticks—may not translate, one-to-one, as a 30 percent lower risk of Lyme disease. A single method may have to be intensive, aggressive, and sustained to achieve the desired reduction of disease risk. This is bad news for those who would put all eggs in one basket. But it's encouraging for those advocating a blend of complementary approaches, which singly may provide only modest effect but in aggregate add up to substantial benefit.[5]

Acaricides

Acaricides kill ticks.[6] Ticks and mites constitute the Acari family of arachnids and are distinguished from spiders. The "-cide" part of the name has the same root as the last syllable in "homicide." Most chemical acaricides are also insecticides, and both acaricides and insecticides fall under the much broader term of "pesticide," which includes chemicals and biological agents with a variety of functions, including herbicides, fungicides, antifouling agents for boats, and rat poisons. All pesticides distributed and sold in the United States must be registered with the Environmental Protection Agency (EPA). Individual states also have to approve pesticide use within their borders. For simplicity here, I use the more familiar term insecticide interchangeably with acaricide. Keep

---

* A particularly good single source at this time is the state of Connecticut's freely available booklet *Tick Management Handbook* (www.ct.gov/caes).

in mind, though, that only a small subset of insecticides are in common use against ticks, and there may in the future be bona fide acaricides that have little or no effect on insects. Some of the insecticides reviewed here are only for application to the land, others are targeted to animals, and another group are used in more than one way.

We commonly think of an insecticide as a manufactured chemical, and typically it is a small molecule produced from scratch in a factory. Some insecticides, like malathion, chlorpyrifos, diazinon, and other organophosphates, are simple in structure and do not otherwise occur in nature. Organophosphates are related to the nerve gas agents used in warfare and act by disrupting an enzyme involved in transmitting nerve impulses. For arthropods exposed to comparatively low doses, this action can be lethal. The few remaining organophosphate insecticides allowed for use in the United States do not persist in the environment or in the bodies of animals, in contrast to DDT and chlordane, two organochlorine insecticides. Organophosphate insecticides are considerably less potent for humans than military nerve agents. However, because of the inherent toxicity of their mode of action and the ease with which inadvertent overexposure can occur, organophospates are generally restricted for residential and other nonagricultural purposes to licensed commercial pest control companies and to government agencies with vector control and mosquito abatement responsibilities.

As nonagricultural use of organophosphates waned, other insecticide classes took their place. One of the most commonly used insecticides now is carbaryl, which has the full name of 1-naphthyl methycarbamate and is commonly known under the trade name Sevin. Carbaryl has a similar inhibitory effect on the nerve impulse enzyme as organophosphates, but it is a carbamate, with carbon, hydrogen, oxygen, and nitrogen atoms but no phosphate. Carbaryl is approved for residential use and for home gardens, as well as for commercial agriculture and forestry and rangeland protection. It has frequently been used as an acaricide, especially by commercial pest control companies. Though carbaryl was judged safe enough for approval for shampoos for pets and has been used for treatment of people with head lice, it is highly toxic to honeybees and to aquatic animals exposed to runoff from farms and lawns. Carbaryl's use in pet flea-and-tick collars was ended by the EPA in 2009. Carbaryl is illegal for any application in several European countries and is more restricted in its use in Canada and Australia than in the United States, where it is classified as a "general use pesticide." (Another insecticide of the carbamate class is propoxur, which is used in some flea-and-tick collars for pets.)

Fipronil is an insecticide that is used against ticks only in animals. For tick-killing purposes, it is not dispensed on the ground and plants. It is a broad-spectrum insecticide of the phenylpyrazole class and causes central nervous system toxicity by directly inhibiting transmission at nerve junctions. Because it is a slow-acting poison, it is particularly useful as the active ingredient in baits for social insects. A fire ant or yellow-jacket wasp is attracted to a bait trap, consumes the bait, and returns to a nest where other occupants are exposed as well, to their misfortune. Another advantage of fipronil for baiting is its lack of repulsiveness for arthropods (and by that criterion it makes a poor repellent!). Negatively affected creatures include bees, aquatic invertebrates, fish, some birds, and insect-eating mammals. Fipronil is one of the insecticides that is implicated in colony collapse disorder of honeybees, and its use is restricted in the European Union. In the United States, it is registered by the EPA for bait treatments and topical (spot-on) treatment of domestic animals.

Amitraz is another synthetic insecticide used in applications tailored for delivery to animals, such as pet collars or wildlife baits. Amitraz, whose full name is the unpronounceable N,N'-[(methylimino)dimethylidyne] di-2,4-xylidine, is a chemical that also works on the nervous system of insects and ticks, in this case by stimulating nerves with the same effect as adrenaline. As a consequence, nervous activity increases, to the arthropod's detriment. Amitraz is as much known for its success against mites, including plant pests, and ticks as it is for its anti-insect applications. According to the EPA, amitraz is generally toxic for aquatic invertebrates and fish and for some birds, but is comparatively nontoxic for bees.

Two other insecticides are increasingly used in products for pets and other domestic animals. One is imidacloprid, a neonicotinoid class insecticide, which blocks transmission in one of the nervous system's pathways, paralyzing the arthropod. Although it is effective against fleas, it does not kill ticks. The second is afoxolaner, which is a member of the isoxazoline class of insecticides and acts on another key site in the transmission of impulses in the nervous system from other insecticides. Unlike the other insecticides reviewed, afoxolaner is administered orally. In one study, when dogs were treated with afoxolaner, ticks on the dogs, including *I. scapularis*, were killed within 48 hours of their attachment.[7]

Insecticides from Plants

Another major group of insecticides, the pyrethroids, trace their origin to a natural product, an extract from the pyrethrum daisy, a type of chrysanthemum. A "pyrethrin" is a pyrethroid that is an unadulterated extract from the plant. Synthetic pyrethroids have a similar structure to the pyrethrins, but they were created in the chemistry lab and have been modified in different ways to increase their stability in the environment, to enhance other desirable qualities, and to decrease toxicity for other animals. One of the oldest and best-known pyrethroids is permethrin. There is a good chance that an insecticide is a pyrethroid if its name ends in "-thrin." Other examples of pyrethroids are allethrin, binfenthrin, cyfluthrin, cyphenothrin, deltamethrin, and tetramethrin. Sometimes a pyrethroid is combined with another compound, such as piperonyl butoxide, to increase its potency.

Permethrin and other pyrethroids, like most other insecticides, act on the nervous system of insects and ticks. This particular class of compounds disrupts the function of the nerve cells themselves. Mammalian nerves can also be affected but not as easily, and mammals' larger size means that insecticide is more likely to be broken down by our body before it reaches our brain—whereas in an insect's body, the action is more immediately potent. Cold-blooded animals like insects and ticks, but also fish and invertebrates, are more susceptible to the effects of pyrethroids than mammals or birds.

Pyrethroids are used as a spray or in granular applications for landscapes and around residences, to coat or impregnate the fabrics of clothes for personal protection, and as ingredients of spot-on preparations and in flea-and-tick collars for dogs (but not cats). They are also approved by the FDA for use by humans as creams and lotions for treatment of head lice and scabies, a skin disease caused by mites.

Pyrethroids are attractive for seemingly being more "natural" than insecticides that are born in a chemistry laboratory. They are akin in this sense to antibiotics (chapter 8), which derive from natural products produced by fungi and bacteria. We generally think of antibiotics as mostly benign for humans, save for the odd side effect or allergy. But some antibiotics were discovered early on to seriously damage cells of mammals and other vertebrates, not just bacteria. Certain of these toxic antibiotics, like daunomycin from a *Streptomyces* bacterium, have been successfully exploited for use in cancer chemotherapy. They kill the proliferating cancer cells at a faster rate than healthy cells. As life saving as these anticancer antibiotics have been, they are restricted from any other use besides cancer therapy, because of their toxicities for normal cells. So beginning with a "natural product" does not always equate with less harm. After all, some of the most potent toxins on Earth are natural products, like botulism toxin from a bacterium, ricin from castor bean plants, and several mushroom poisons. Moreover, even if permethrin and especially the newer synthetic pyrethroids are safer for people than organophosphate or carbamate insecticides, they can still adversely affect nontarget species like honeybees, aquatic invertebrates, and fish. Cats tend to be more sensitive to the effects of pyrethroids than other mammals, and for this reason there are no flea-and-tick collars with a pyrethroid available for cats. Permethrin has been banned by the European Union since 2011.

As more restrictions, if not outright bans, are placed on traditional insecticides like organophosphates and carbamates, there is increasing interest in other plant-derived compounds. (Why plants? Plants are at the front lines of the battle with insects and mites and would likely have adapted with chemical defenses of their own.) These plant-derived insecticides include, for example, rotenone and products based on rosemary oil. Another trend driving the search for alternative insecticides is increasing resistance to the existing agents among insects and ticks. Acaricide resistance is not now so much a problem in *Ixodes* species that transmit Lyme disease as it is for ticks of agricultural importance, such as cattle ticks, which have been heavily exposed to insecticides over decades in places like Texas, Australia, and Brazil, where tick infestations of cattle have an economic impact.

Area-Wide Acaricides

A common method of tick control for many years has been to apply an insecticide directly to a defined space, such as a residential property, or even to a larger area. Indiscriminate dispersal of an acaricide over an entire region by terrestrial or aerial spraying without doubt kills a lot of ticks (as well as many other arthropods and invertebrates). Widespread application of the insecticide DDT by aircraft reduced the risk of tick-borne encephalitis in the former Soviet Union.[8] Needless to say, the application of insecticides on this scale would not be tolerated in most of North America or Europe, due to public concerns about toxicity and the effects on nontarget organisms.[9]

To minimize needless exposures of people and other animals, the spraying of liquid or dispensing of granular forms of insecticide is targeted to microenvironments where ticks are most likely to be on the property, such as where the lawn meets the woods, ornamental ground cover, stone walls, and several meters into adjoining woodlands.[10] This way the majority of the lawn itself and herb and vegetable gardens can be insecticide-free. The timing of the application is also important. Since the nymphal ticks pose the biggest danger by far, insecticides are best applied in the late spring or early summer, when this stage of ticks is active. To kill the adult ticks, thus limiting the supply of immature ticks the next spring, insecticide can be applied in the fall. These periods apply to the northeastern and much of the north-central United States. The optimum times for the annual insecticide application may be different in other regions, such as Northern California.[11]

Acaricides can be applied by the homeowner but it usually is done by a commercial pest control, arborist, or landscape care company, which are licensed and registered for use of insecticides. Licensed operators may have access to insecticides that are otherwise not available to homeowners and other nonprofessionals. Whoever applies the insecticide, there are some common sense guidelines to follow:

1. Don't use if it's windy (more than 10 miles per hour).
2. Close the windows and doors before spraying or dispersal.
3. Keep away from wetlands, bodies of water, and wellheads.
4. Avoid areas where there are flowering plants that attract bees and other pollinators.
5. Keep household members and pets, especially cats, off the treated area for 12 to 24 hours, or longer if indicated.
6. Don't water the lawn or apply within 24 hours of rain, to minimize runoff.
7. Check to see if your state or municipality requires notification of neighbors or posting of signs for urban or suburban applications of insecticides.

A single well-timed annual treatment with an acaricide in this selective way can reduce populations of nymphal ticks by 90 percent or more. But less than 25 percent of residents, even in high-risk Lyme disease areas, practice this method of prevention. As we'll see, there are other options for attacking ticks in the landscape, but the gold standard remains area-wide application of an acaricide.

Biological Control

An alternative control measure to insecticides is a living predator or parasite that is put to service.[12] A well-known biological control agent that you can buy at the nursery is a container of ladybugs for your aphid problem in the garden. But finding such a natural enemy of ticks is not easy. Ticks feed only on blood, and then only once a year. They are also solitary, seldom close to other ticks for most of their lives. There are birds that are natural predators of ticks, but they are native to Africa, where they eat ticks on grazing ungulates, like zebras. Domestic fowl, including free-ranging chickens, find and eat ticks if given the opportunity, but they would not be expected to have much overall impact.

What seemed a more promising candidate for biological control was a wasp (*Ixodiphagus hookeri*) that lays its eggs in ticks.* The tick itself is the unfortunate host. After the female wasp deposits its eggs inside the tick, the immature insects begin to grow inside, devouring it from the inside like the creatures of the *Alien* movies. These and other wasps with a similar lifestyle are called "parasitoids," because they parasitize other animals but only for part of their life cycle. When deer ticks are at very high numbers in an area, local release of the wasps can reduce tick densities in the succeeding year. But as time goes on, when there are fewer ticks for the wasps to parasitize, the wasp populations disappear or decline.[13] So unless huge numbers of wasps were to be released every year, the long-term success of this approach is doubtful. Even if a yearly release was chosen, there would need to be improvements in the artificial rearing of the wasps to make this approach cost effective.

Another proposed candidate for achieving biological control are tiny round worms that go after ticks.[14] Most animals in the world, including humans, carry some type of worm, so it is not surprising that there are worms that specialize in ticks as hosts for one of their life stages. It isn't the worm itself that damages the tick. Certain bacteria that the worms carry do the dirty work. Although an effect of the worms on *I. scapularis* has been shown in the laboratory, there are challenges for this application to be effective in the field. For one, the worms mainly target engorged adult female ticks that have dropped off deer to the ground. That's good, one would think: the next generation of ticks would be diminished. But only one stage of the worm's life cycle uses ticks as a host; the hosts for other stages are not in the same part of the United States. The worms could not sustain themselves in the environment. Consequently, controlling ticks would require dispensing the worms repeatedly to the same area.

Somewhere on the size scale between an insecticide molecule and a wasp observable with the naked eye is biological control with a microbe, specifically a fungus or bacterium that attacks the arthropod pest. These

---

*The genus name *Ixodiphagus* suggests that only *Ixodes* species ticks are "eaten," but the wasp also deposits its eggs in other types of ticks, some of which are no threats to people.

are live organisms that are applied like a chemical. A bacterium called *Bacillus thuringiensis* (Bt) is widely and successfully used as a control agent for a variety of pest insects in commercial agriculture, as an alternative to insecticides for mosquito control, and even for insect control by home gardeners. The bacterium produces a hardy spore that resists drying and ultraviolet light, so it can be applied as a spray to crops, gardens, or pools of water and left to have an effect later. For the control of insect pests, the spores need to be taken in internally. The spores are like seeds that sprout once inside the insect. The bacteria grow in numbers and make a toxin that kills the arthropod. But how do you get a tick to eat the spores? Ticks only eat blood, and spores would not be found in the blood of animals. There is a report of an effect of *B. thuringiensis* on deer ticks in the laboratory,[15] but how the bacteria gain entry inside the ticks is not known.

Better candidates than bacteria for biological control are fungi that attack ticks. The candidate species, *Beauveria bassiana* and *Metarhizium anisopliae*, are part of a larger group that includes bread mold, baker's yeast, and truffles, among others. Unlike *B. thuringiensis*, the fungi do not need to be consumed to do damage. The fungus can penetrate the tough shell of the tick from the outside, and then it grows, like mold on a tomato. Major obstacles to the practical application of fungi for biological control are how to produce the fungal spores in large quantities and how to deliver the spores to the ticks in a cost-effective way. The fungi are living organisms and can be affected by environmental conditions, such as humidity, as well. The fungi that attack ticks may also be pathogenic for other types of arthropods, so there may be deleterious effects on nontarget species.

Pheromone-Based Agents

Pheromones are fancifully the natural equivalent of perfume and after-shave lotion for animals. They are small molecule chemicals that include sex attractants for mating and the means for signaling to gather in assemblies. Some pheromones have effects in a variety of species, while others are more genus- or species-specific. Tick pheromones act similarly whether they are functioning as sex attractants or for assembly and might be exploited for control in various ways. These include luring ticks to an acaricide or a biological control agent, like a fungus, either on the ground[16] or on an animal wearing an impregnated tag or collar.[17] Sex attractant pheromones would target adult male ticks, since they are the ones tasked to locate a female. Although male ticks are seldom the main concern for infection transmission, they are indispensable for reproduction of the species. If there are few or no male ticks on a deer or other host, no matter how many females are about, the number of ticks in the next generation will be few to nil. Most of the research has been done on tick genera other than *Ixodes*, though, so we wait to see how sex attractant pheromones succeed on other testing grounds, such as for cattle ticks on ranches or for dog ticks around kennels.

Targeting the White-Footed Mouse

Monhegan, an island off Maine, was an ideal place to investigate the contribution of a small rodent host for *B. burgdorferi* and *I. scapularis* to the Lyme disease problem.[18] The Norway rat, an urban denizen throughout the world, was an invasive species on the island. The rats served as reservoirs for *B. burgdorferi* and hosts for *I. scapularis* ticks in place of the absent white-footed mice. This rat is little loved, and consequently the community gave its approval for rat control measures. Poison put out for the rats was effective in lowering the rat population on the island. However, this triumph came at a price. A resident's pet, a terrier fond of hunting rats, became very sick after eating a poisoned rat carcass. After this incident, it was difficult to gain approval from residents for this manner of rat control, even when the rat was an unwanted invader.

A place with rats instead of white-footed mice was a useful accident, but an accident of history nonetheless. What about intentionally eliminating *Peromyscus leucopus*? This is feasible on an island but harder to pull off on the mainland, where there would be incursions by mice at the edge of the elimination zone. But even on an island, eradication of a small mammal, like *P. leucopus*, would be logistically challenging (expensive) and contentious. If there was controversy over eradicating the Norway rat, the potential carrier of plague and all-around household pest, imagine what might happen if the animal in the exterminator's sights was the more adorable and otherwise harmless white-footed mouse? But we can get some idea of the effect of this strategy by examining other islands where the white-footed mouse is already absent. One such is Patience Island of Narragansett Bay, Rhode Island, which has meadow voles (*Microtus pennsylvanicus*) but not *P. leucopus*. And guess what? The voles serve about as well as hosts for immature *I. scapularis* and as reservoirs for *B. burgdorferi* as do the mice.[19] There are other potential hosts for the ticks and the pathogen to take the place of voles should humans decide to go after voles too. Pretty soon there would be no furry little animals on the island, an undesirable outcome for most sensibilities. (Birds are an alternative target, but reduction of the numbers of foraging birds would likely only have an effect in Europe, where the bird-dependent species *B. garinii* occurs. And, needless to say, trying to eliminate birds from an island a few miles from the mainland would be futile.)

Another approach targets *P. leucopus* (or other small mammals) not to lower their numbers but more benignly as means to get at the ticks. This strategy better fits with our view of the tick as the "pump handle" for Lyme disease. There are two ways to entice mice to come to the intervention. The first takes advantage of the rodent's nest-building behavior and adds an insecticide to some material that a mouse would incorporate into a nest.[20] The insecticide rubs onto the rodent's fur and thereafter functions to thwart the attachment of ticks. This theory was put into practice with paper tubes containing permethrin-impregnated cotton; the commercial product is called Damminix. The idea is that property owners or parkland managers spread the tubes out on the lawn and forest on their land. The larger the property, the more tubes are needed. As appealing as this idea is, it worked in only one of four experimental studies. On an island off the coast of Massachusetts, distribution of the tubes did reduce the numbers of ticks in the area, but in three other studies carried out at mainland sites in New York and Connecticut, which may have had a larger variety of alternate hosts than on the island, there was not a significant effect.[21]

Food is the second enticement for targeted insecticide application to rodents. Attracted by the bait's odors, white-footed mice, chipmunks, or other rodents enter tubes or boxes that are narrow enough for the animal's fur to come in contact with an insecticide that coats the interior. Trials of this method were carried out in the Northeast[22] and California.[23] They were moderately successful. From 2004 to 2006, a commercial product called the Maxforce Tick Management System, based on the original prototype, was available, but it was pulled from the market by the EPA. One drawback of that earlier device was that squirrels and raccoons were attracted to the boxes and, in trying to get to the bait, destroyed the tubes. The devices were also not sufficiently child resistant. A second-generation bait box, called the Select Tick Control System, was introduced in 2011. The insecticide is a low concentration of fipronil. The devices have access ports that are large enough for the important reservoirs, like white-footed mice, chipmunks, voles, and shrews, but that have metal shrouds to prevent depredation by squirrels and raccoons. The EPA and at least thirty-two individual states have registered these bait boxes, but only for use on residential properties by licensed professional pest management companies.

Targeting the Pathogen in Reservoirs

Administration of a vaccine to white-footed mice is a plausible strategy for Lyme disease control.[24] The near-term objective is to reduce the density of infected nymphal *I. scapularis* ticks in the environment. The density of infected nymphs is a good indicator of the risk of Lyme disease for humans in an area. The rationale behind this vaccine approach is this: antibodies in the blood of an immunized animal kill the spirochetes in the intestine of the tick as it feeds on the mouse. This occurs when, one, the tick had previously been infected as larva and is now feeding as a nymph or, two, an uninfected larva feeds on an infected but immunized mouse and the spirochetes enter the intestine together with the antibodies. The latter phenomenon sounds paradoxical. If the mouse was infected despite the immune response to the natural infection, why would the spirochetes be killed when they entered the tick with blood that seem-ingly was ineffectual?

The apparent paradox is explained by the change in the spirochete's surface proteins when it goes from a mammal to a tick (chapter 3). If the vaccine is based on a tick-specific protein—that is, one that is not pro-duced by the spirochete while inside a mammal or bird—then even if the spirochetes have managed to evade immunity to that point, they become vulnerable after they move from the mouse to a tick. When the bacteria switch to making the new protein inside the tick, the vaccine-elicited antibodies, which had been taken on board with the blood, pounce when they recognize cells making the protein. Think of it as a Trojan horse tactic. The flip side is that vaccination of an infected mouse with a tick-specific protein will be of no benefit to the mouse in clearing an infection under way, because the spirochetes are not making the vaccine protein. But that mouse would be protected against a second or third infection with *B. burgdorferi*.

The beneficial effects of the vaccine are threefold and cumulative: First, an immunized, previously uninfected mouse is protected against infec-tion, thereby reducing the total number of infected mice by one. Second, an uninfected larva that feeds on an infected mouse fails to pick up *B. burgdorferi*, leading to one less infected nymph to bite a human. And third, a previously infected nymph is cured of the infection, resulting in one less infected adult. The last effect would lower Lyme disease risk mar-ginally, because relatively few human infections are acquired from adult ticks. But the first and second effects could result in a substantial lower-ing of the number of infected nymphs that people are exposed to in the succeeding years.

The idea of vaccinating wildlife is not farfetched. This has been the policy and practice for many years for the control of rabies by public health agencies.[25] Wild animals, such as raccoons and coyotes in North America and red foxes in Europe, receive an oral rabies vaccine in food bait. This bait is distributed in forests and other undeveloped areas where rabies in wildlife is a threat. The vaccine is a modified version of the human vaccine against smallpox and has been engineered to make a protein of the rabies virus. This type of vaccine can also be adapted for eliciting immunity against a *B. burgdorferi* protein.[26]

So why not proceed and vaccinate white-footed mice? There are two cautions. The first is the aforementioned fact that there are other hosts for *B. burgdorferi* besides *P. leucopus* in the northeastern and north-central United States; these include shrews and chipmunks, among others. Extending immunization to other animal species beside *P. leucopus* complicates the prevention program, especially if you need to tailor the bait and delivery systems for each species. The local shrews, which are not rodents, would be particularly hard to reach, because their diet is mainly carnivorous. And then there are the additional regulatory requirements from government agencies, like the EPA and the U.S. Department of Agriculture, as target species add up. Proposals for further interference with wildlife may be skeptically viewed as undue meddling in the environment.

The second issue is the increasing threats from other infectious diseases, such as babesiosis, HGA, relapsing fever, and deer tick virus encephalitis (chapter 11), that the deer ticks carry. There may be a lower frequency of *B. burgdorferi* in nymphs as the result of a vaccine campaign zeroing in on that pathogen, but that would not make a dent in the risk from other pathogens. Development of vaccines that are effective against more than one pathogen is conceivable and could take the form of either a combination of individual vaccines or an antitick vaccine that blocked uptake of more than one agent. But unlike a vaccine against *B. burgdorferi*, which could be based on vaccines approved for use in humans and dogs (chapter 13), there is not much in the vaccine development pipeline for babesiosis, HGA, relapsing fever, or deer tick virus encephalitis.

A potentially cost-effective alternative to vaccination of wildlife is mass administration of an antibiotic in food bait. In a two-year study of the antibiotic doxycycline delivered in bait to small mammals, there was not only a substantial drop in the infection rates for *B. burgdorferi* and *A. phagocytophilum*, the cause of HGA, in the rodents, but also more than 90 percent reductions in the carriage rates of these two pathogens in the nymphs.[27] Although this is an impressive outcome and an encouraging proof of principle, widespread implementation of this intervention risks selecting for antibiotic-resistant mutants of the pathogens. As discussed in chapter 8, a newly arisen antibiotic-resistant mutant in a human would be at a dead end; it has nowhere to go. But that would not be the case for a doxycycline-resistant *B. burgdorferi* that might be selected for in a treated white-footed mouse. The mutant could be spread by ticks and then proliferate in other mice with the antibiotic in their bodies. Loss of doxycycline as a therapeutic option for *B. burgdorferi* or *A. phagocytophilum* in humans would be highly regrettable.

Targeting Deer, the Host for Adult Ticks

Previous chapters have described the abundance—some would argue overabundance[28]—of white-tailed deer in much of the United States and Canada. The vignettes about Nantucket and Block Island illustrate the impact that the proliferation of this species can have on individual communities. Deer population sizes are a relevant consideration in areas where *B. burgdorferi*, *B. microti*, *A. phagocytophilum*, *B. miyamotoi*, and tick-borne viruses occur, because the vector for these pathogens is the deer tick, *I. scapularis*. One study estimated that in the northeastern United States, 95 percent of adult female *I. scapularis* fed on white-tailed deer, with raccoons a distant second at 4 percent.[29] Other areas in the United States are affected by different diseases transmitted by other types of ticks, like the Lone Star tick, a major host for which is the white-tailed deer.[30] For this reason, other localities and health and environmental agencies have been contending with the effects of large deer populations as well. Some of the methods used in deer in Lyme disease country are adapted from those evaluated and proven useful elsewhere.

This is not the place to consider at length the pros and cons of white-tailed deer populations as they exist today in North America.[31] The tangible and intangible benefits, such as opportunities for hunting and the pleasures of having these beautiful creatures around, are weighed against perceived costs, such as deer-vehicle collisions, negative ecologic impacts on other species, and damage to agriculture, the timber industry, and residential and commercial properties. How one values each of these pleasures and problems affects opinion about Lyme disease–prevention measures that are directed at deer.

Let's start at one end of the spectrum of interventions: remove deer entirely from a community. As we saw, elimination of a single species of small animal host for immature ticks would have limited effect on either the number of *I. scapularis* or the prevalence of *B. burgdorferi*. Adult ticks too will take to another animal, if that's all that is available. But for reproductive success of the population over time—and that's what counts—the host really needs to be a large enough animal for there to be female and male adults present at the same time. As a single measure, eliminating deer would be expected to have a larger effect than eliminating one species of small animal hosts for immature ticks. So it's logical—if not politically astute—to propose this.

We cannot turn back the clock to the first part of the twentieth century, when deer populations were at their nadir, and at that point set up of controlled prospective experiments, where some areas were allowed to have deer without restriction and other areas were prohibited from having deer at all or from having more than small numbers. That horse—or deer—has left the barn, so to speak. We can try look at some isolated instances where deer populations have been manipulated during the Lyme disease era and see what happened. Not surprisingly, given the contentiousness over moderate culling of deer herds, let alone complete removal, there have been only a few examples to study.

One instructive example is Monhegan, that island off the coast of Maine.[32] White-tailed deer were introduced to the island in 1955, and by the mid-1990s, the population had grown to a density of thirty-seven per square kilometer, a value comparable to what was on some of the other islands. Lyme disease was frequent among residents and visitors on the island, but in other respects Monhegan was not particularly representative of the areas in the Northeast where Lyme disease is a risk. This is where there were Norway rats instead of white-footed mice, and the only alternative hosts for adult *I. scapularis* were humans and their dogs and cats. Notwithstanding these limitations, we can learn from the 1996–1999 program to remove the island's deer. The investigators used historical data from Monhegan, as well as a reference island without intervention, to assess the effect, the primary measures of which were counts of immature and adult ticks on the land and the infection prevalence in the collected ticks. For the collection of ticks in fall 1999, the densities of adult tick went up, probably because they were still actively looking for hosts. But thereafter tick counts and infection prevalences fell. By the summer of 2003, no immature ticks were found on captured rats, and there was more than a tenfold fall in the density of adult ticks. The absence of deer could be sustained, of course, because this was an island 10 miles from the coast, too far for even the most adventuresome buck or doe to swim.

Studies on the mainland, which were carried out under more representative ecologic conditions, recorded similarly lower tick densities at test sites from which deer were excluded. But these exclusions required 8- to 10-foot-high fences or electric fences keeping the deer out. Moreover, the testing grounds were relatively small plots, some of which were on private lands. A more realistic goal for a community-wide intervention is reduction, not elimination, of the local deer population. This may be justifiable to the public and its elected officials not just on disease prevention grounds but as a means to mitigate residential and agricultural damage by deer.

There are various ways to partially cull a deer population, some more feasible or acceptable than others. For instance, reintroduction of predators, such as wolves, would undoubtedly have the desired impact, but this is not likely to win approval by a township in the northeastern United States. The deer can be humanely trapped, but what other community would want them? What about contraception for the deer, seemingly a nonbrutal tactic for reducing deer herd sizes? There are methods of contraception for deer, including contraceptive vaccines, which would not entail use of a drug in a food animal.[33] But deer contraception requires injecting individual deer and may be more effective on islands where entry of outside deer can be restricted. Finally, there is the option of trap-and-kill or bait-and-kill using professional sharpshooters or by lottery to hunters. This may be the only plausible solution for many deer within the borders of towns and suburbs and is the approach that is reportedly under way on Block Island, off of Rhode Island, for a controlled reduction in the deer herd there.[34]

But how much of a cull is enough to make a difference in disease risk? There is an overall direct correlation between the number of deer in an area and the number of *I. scapularis*, but this is with considerable year-to-year fluctuation in the number of ticks. Clearly, there are other factors, such as weather and availability of hosts for larval and nymphal ticks. This means that assessing the effect of incremental deer reduction on a short timescale may be a challenge. Limited experiments in the field and computer modeling indicate that the reduction would need to be substantial, up to 90 percent, to make a significant difference in the figure that counts most—the density of nymphal ticks—in succeeding years. There might even be a short-term increase in the risk of Lyme disease after a partial cull, as there may be more adult ticks available for biting people, and other infected adult ticks may feed on smaller mammals. More experimental field studies with appropriate controls, such as similar areas without a cull, are needed.

For a community that has reduced its deer population in the immediate area and wishes to limit influx of deer, an option is deer exclusion by high fencing. (Deer can leap as high as 10 feet, or 3 meters.) As described above, fencing can be effective on a community-wide level or for large estates or collections of residences in reducing the number of immature ticks within the fenced area. But the size of fenced area makes a difference, because immature ticks can be reintroduced by small mammals and birds at the periphery. There may be fewer ticks, but those that are present within the enclosure may still be disease carriers if they have fed as larvae on an infected animal.

Using Deer as an Attractor of Ticks

Ticks, lots of them, congregate on the skin of deer. If deer can be enticed to a certain location with food or a salt lick as bait, they can then be doused or brushed with an acaricide. What is called a "passive topical treatment" provides a way to attack mature *I. scapularis* ticks in this concentrated way, thereby limiting tick reproduction and eventually reducing the entire tick population in the area.

A particularly efficacious method for achieving this employs the "4-poster" device, which was first developed by the U.S. Department of Agriculture for controlling the Lone Star tick, *A. americanum*.[35] A central bin, the size of a garbage can, stores and dispenses bait, usually whole kernel corn, to attract deer. As up to two deer try to get at the food at each station, the animals rub against acaricide-impregnated rollers on the posts, thereby receiving the acaricide directly to the head, ears, and neck. The acaricide is then transferred to other body areas by self-grooming. The acaricides are either permethrin or amitraz, both of which are effective against ticks, are comparatively nontoxic, and have long residual activity on the fur of deer. To be effective in the northeastern United States, the minimum density seems to be at least one device per 50 acres (20 hectares). A review of the outcome of the 4-poster implementation at several sites in the northeastern United States showed that after six years of treatment, the density of *I. scapularis* nymphs was reduced on average 71 percent in comparison to similar control sites.[36]

In 2012 the state of New York allowed the use of 4-poster stations dispensing permethrin to certain counties with a high incidence of Lyme disease. One of the first places to test the devices was Shelter Island at the tip of Long Island. According to a news report, Shelter Islanders had lobbied for this allowance from the state government for more than a decade.[37] The town continued to deploy fifteen 4-poster stations at an annual cost of $75,000.

An alternative to topical delivery of an insecticide to deer at a bait station is provision of a systemic acaricide in the food itself. This is usually a member of another class of compounds, the macrocyclic lactones, which are routinely used as oral drugs to treat and prevent worm infections of domestic animals and people. The best-known member of this class is ivermectin, which in oral form is used to prevent heartworm in dogs; another class member used in products is selamectin. The drawbacks of this approach are the medicine's cost at the scale it would be needed for an animal the size of a deer and the risk of overdosage. Because deer are food animals, there are also concerns about the presence of these chemicals in their meat, which may be consumed by humans. To be effective against adult ticks, the deer would need to be treated in the fall, just when hunting season gets under way.

Antitick Vaccine for Deer

If animals can be attracted by bait to a location where they can be passively doused with an acaricide or treated with a medicine, they also could be administered a vaccine—if it could be orally delivered. A vaccine of this sort would immunize deer against ticks and, if successful, would offer an alternative to pesticides for reducing tick populations. Biologists know that if animals are bitten enough times by ticks, some will develop immunity to them, to the extent that the ticks fall off an immune animal sooner than they would from an animal naïve to ticks. A group of scientists in Australia took this natural phenomenon a step further by immunizing cattle against proteins of the tick intestines.[38] When a tick feeds on one of the vaccinated cows, the antibodies in the animal's blood damage the tick's intestine, thus killing it. A commercial vaccine based on this principle is in use for cattle in different countries. The problem for would-be developers of a deer vaccine is that the cattle vaccine has to be administered by injection. There is to date no oral formulation. It is one thing to inject domesticated animals like cattle, which are used to being rounded up and in confined quarters; it is another matter to track down and administer shots to deer in forests. It is plausible to do this— as it is plausible to administer contraceptive vaccines to deer—but it would be labor intensive and expensive. And, if the experience with the cattle vaccine is a guide, the deer may need to be reinoculated every six months.

The Effect of Biodiversity

As might be clear by now, the ecology of Lyme disease agents is complex. There are multiple potential reservoirs for the pathogen. The tick vectors are not specialists; they feed on different hosts at different stages of the life cycle. The weather, the presence of predators of the various hosts, the

availability and nutritional value of food for the hosts, and other parasites all enter into the equation. One theory, which incorporates some
of these different variables, holds that increasing biodiversity, specifically
diversity of the vertebrate hosts for the pathogen and tick, would reduce
pathogen transmission among wildlife.[39] If there were a greater range and
number of alternate vertebrate hosts for the immature ticks, these could
serve to dilute the contribution of *P. leucopus* to the total number of
nymphs. If these alternate hosts were also less suited for the growth of
B. burgdorferi than *P. leucopus* is, then the consequence of the dilution
effect would be a lower prevalence of *B. burgdorferi* in the nymphs, thereby
lowering risk of Lyme disease for people. Although increasing or at least
maintaining biodiversity may be justifiable on several grounds, not the
least of which is the irreversibility of extinction, maximizing the number of potential host species per se may not reduce Lyme disease risk; indeed, the risk could increase under some scenarios.[40] Moreover, if there
are still a lot of white-footed mice about, that may trump the reintroduction of another species or two.[41]

Combining Disease Control Strategies

What I referred to at the top of the chapter as a "blend of complementary approaches" for achieving reduction in disease risk has a name:
integrated pest management. According the EPA's broadly serving
definition, integrated pest management (IPM) is "an effective and
environmentally sensitive approach to pest management that relies on a
combination of common-sense practices."[42] As IPM is practiced for agriculture, economic considerations are heavily weighted. The value of a
head of cattle at a certain age and in a certain place is known. Putting a
dollar value on a Lyme disease case that is prevented may be impossible,
but equipment and materials costs and personnel time are nonetheless
important determinants for households, commercial operators, and public health and vector control agencies. One example of a successful combined approach for reducing Lyme disease risk was a two-year study in a
suburban residential area in New Jersey. The local health department and

CDC investigators used three methods with three different acaricides and three different ways to deliver them: (1) application of a granular pyrethroid acaricide to boundaries between forest and lawn of the participants' properties; (2) 4-poster stations for topical application of amitraz to affect adult *I. scapularis* on deer; and (3) food bait boxes for topical application of fipronil to small mammals to affect immature ticks. After two years, the populations of deer tick larvae, nymphs, and adults were each reduced by about 90 percent.[43]

That's what I call removing the pump handle!

Chapter 13

Preventing Lyme Disease
Personal and Household Protection

People traveling to countries with malaria can decrease the chances of getting infected by taking personal precautions. A nearly foolproof precaution is to steer clear of those regions of the country where malaria transmission occurs. The rural areas may be malarious, but the urban centers are not. The low-lying coastal area may be risky, but the highlands are malaria-free. If vacation or work plans include those riskier areas, there are window screens and bed nets to keep those pesky and potentially disease-carrying mosquitoes at a distance. If outside activities are on the schedule, malaria precautions involve covering up with a hat and lightweight but close-weaved clothes and applying an insect repellent to exposed skin. Another preventive measure, or prophylaxis, is taking an oral antimalarial drug during the trip and for a time thereafter. If those measures fall short and a person gets malaria, there are antiparasitic treatments available. But vaccination against malaria is not currently an option for either visitors or residents. For Lyme disease, we can draw parallels with each of these measures.

Avoiding Ticks

Incidence maps for Lyme disease, which are available on the Internet from the Centers for Disease Control and Prevention (CDC) and state health departments in the United States, provide residents and visitors a satellite's eye view of large areas where Lyme disease or other tick-borne infections occur. Lyme disease maps are comparable in detail to maps of malaria risk for travelers. But they are not detailed enough to discriminate between higher and lower risk environments within those large regions. The distribution of infected ticks can be highly focal. A high-risk suburban housing area or parkland may be adjacent to a lower risk one. We need a crowd-sourcing site, where people can report tick encounters with automated GPS coordinates on an app, like Google's Waze for reporting traffic slowdowns. This would provide more detailed, real-time information for assessing the exposure risk for ticks on a map. One could then hopscotch between low-risk areas if need be. But we're not there yet.

The forewarned resident, worker, or visitor gains useful practical information by learning what the disease-carrying ticks look like. There are many sources of pictures on the Internet of various kinds of ticks (see the Internet sites listed after this chapter). Field guides in libraries and bookstores are other sources to use in distinguishing one type of tick from another. In the northeastern United States the ticks most likely to be encountered on humans and their companion animals are *Ixodes scapularis*, the Lone Star tick *Amblyomma americanum*, and the American dog tick *Dermacentor variabilis*. In the north-central states, there is *I. scapularis* and the American dog tick but not, as yet, much of the Lone Star tick. In the far western United States, *I. pacificus* is distinguishable from the American dog tick and the Pacific Coast tick (*Dermacentor occidentalis*). *I. scapularis* also occurs in the southeast and south-central states, but the southern form of the species seldom bites humans and is little seen (chapter 3). The brown dog tick, *Rhipicephalus sanguineus*, occurs throughout the world and mainly feeds on dogs, but humans are sometimes bitten. A caveat for tick identification is that species are more readily distinguished in their adult stages. Lyme disease is mainly transmitted by nymphs, which are smaller and less easily distinguishable from other ticks than adult *I. scapularis* females and males, the stages often featured in field guides.

Unlike mosquitoes, which can feed through some fabrics with their long proboscis, ticks need to plant themselves on the skin surface for their meal. Denying them direct skin access impedes transmission. Long-sleeved shirts, buttoned up, and long pants with the bottoms tucked into long socks or sealed with tape, cyclists' straps, or rubber bands will provide this barrier. So will closed-toed shoes. The lighter the color of your clothing, the easier it is to see ticks crawling on the fabric. When entomologists collect ticks in the wild, they often use a white fabric dragged over the ground and low shrubs and bushes. Using a cloth that was black or another dark color would make the entomologists' task very difficult.

When entering landscapes that deer ticks favor, keep to the center of trails to minimize contact with brush and tall grasses. Most ticks attach first to pants, socks, and shoes, unless, of course, a person is sitting or lying down. Once home or back at the hotel, cabin, or camp, check skin and hair for ticks. This is most easily and quickly done by another person, but a "tick check" can also be performed alone, with the aid of a mirror and a fine-tooth comb to locate ticks in the hair. If you are going to shower, do it soon after coming in from outdoors. When you return home, wash your clothes and then put them in the dryer at the highest permissible temperature setting.

Educational programs on self-protection techniques are useful, but given human nature and information overload, their influence on behavior wanes as time goes on. Education is most effectively delivered just before exposure, when people are motivated to attend to the lesson. An example of a successful education program was one offered to ferry passengers as they traveled to Nantucket Island from the Massachusetts coast.[1] Among those passengers who stayed on the island for more than two weeks, there was a 60 percent lower risk of a tick-borne illness for those who received the self-protection program compared with the control group, which received a bicycle safety program instead.

Removing an Attached Tick

As we saw in chapter 3, ticks are well-equipped specialists for attaching themselves to you, digging a hole to sup blood and tissue fluid, and then holding fast for a week or so. A close-up view of the process shows why it can be so difficult to remove that little tick embedded in your skin.[2] Using a mouth with three moving parts, the tick first pierces the skin with the hypostome, a flat probe with barbs on the sides. Then paired appendages, positioned on each side of the hypostome and bearing teeth-like edges, employ a ratcheting mechanism to pull the hypostome deeper into the skin tissue. Once positioned for feeding, the tick keeps tightly in place with the barbs and teeth of the mouth parts as well as the cement-like material it secretes. Abruptly pulling on the tick's body may decapitate it, with the mouth and head parts left firmly stuck in the skin.

If you find a tick already embedded on you or another person or your pet, take a deep breath and carry on calmly. Most of the pathogens, including *Borrelia burgdorferi*, are unlikely to move from the tick to the person for the first thirty-six hours of the tick's attachment. It may even be prudent to wait for the proper tick removal equipment rather than to attempt to disengage an engorged tick with the fingers or a pocket knife. The best tool for tick extraction is a forceps or tweezers. Choose unrasped fine-pointed tweezers whose tips align tightly when pressed firmly together. The points should be sharp enough to grasp the tick's head parts as close to the skin as possible. Do not grasp the tick by the body. Once the points are positioned thus, apply slow, steady traction in a backward direction, away from the skin. Take care not to twist or jerk the tick, to avoid ripping the body off, leaving the head behind. If this happens—and sometimes it does, even with good technique—there is probably little harm in leaving the head part there, as if it were a shallow splinter. In a matter of days, the head part will be pushed up as new skin is formed and the top layers are shed. Trying to remove the embedded tick parts with a knife or other sharp object could lead to skin scarring or introduce infection. A link to a video demonstrating proper tick removal can be found at www.aldf.com/videos.shtml.

An intact extracted tick is probably still alive. Kill it by putting it in ethanol (80 proof cheap vodka will do) or rubbing alcohol. Or simply put it in a tightly capped vial or small bottle—but not an envelope unless it's dead. Trying to crush a still-flat tick may be futile, and it could escape. Do you want a hungry tick loose on the premises?

The site of the bite should be washed with soap and warm water or antiseptic solution after the tick is removed. An antibiotic ointment might prevent infection from other bacteria but would likely have little effect on *B. burgdorferi*. The spirochete is not affected by most antibiotics, like neomycin, contained in over-the-counter wound ointments. Ticks themselves carry few bacteria, but in entering the skin, they may have introduced bacteria, like staph or strep living on top of the skin, to a deeper, more vulnerable layer. An increasing area of redness and swelling around the bite site may need attention. If it occurs within a day or two, it could be a staph or strep infection. If an enlarging red patch appears during the period of 3 to 30 days postextraction, it could be erythema migrans.

A largely ineffective and discredited technique is smearing petroleum jelly or nail polish on the engorged body sticking out of the skin. The theory was that the coating deprives the tick of oxygen, smothering it until it withdraws from the skin to get air. As simple and painless as this approach sounds, it is not recommended. For one thing, it may take several hours to overnight for the tick to be affected. During this time transmission of the infection might occur. This is what you are trying to prevent!

Should the tick be saved? Little is gained from this in most cases, especially if the tick is recognized as one of *Ixodes* species known to transmit Lyme disease. Removed ticks can be analyzed in the laboratory for the presence of *B. burgdorferi*, but doing so requires specialized expertise and equipment, and only a few research centers in the country are capable of doing it accurately. In any case, the expected benefit from prophylaxis (see below) is greater when an antibiotic dose is given soon after removing the tick. A better reason for saving the tick is uncertainty about what type it is. The state or county health department or the local vector control agency can provide advice on where to bring the tick for identification if they cannot do it themselves.

Antibiotic Treatment of Tick Bites

Even in high-risk places in North America and Europe, the overall chance of getting symptomatic Lyme disease after being bitten by an *I. scapularis* or *I. ricinus* tick is about one or two out of one hundred—1 or 2 percent. The odds may go up in areas with particularly heavy infections in nymphs, but they probably are not more than 5 percent per bite. Still, that is not an inconsequential hazard, so it's reasonable to ask whether treatment, or more accurately, prophylaxis, with an antibiotic will prevent Lyme disease from occurring after a tick bite. The potential benefit of preventing illness is weighed against the possible side effects of the therapy (chapter 9).

For North America the answer to the question is yes when these four conditions are met: (1) the tick is identified as a nymphal or adult *I. scapularis* that is estimated to have been embedded for 36 hours or more; (2) *B. burgdorferi* is common in the area, that is, in 20 percent or more of the local deer ticks; (3) prophylaxis is started within three days of removing the tick; and (4) doxycycline or other tetracycline administration is not contraindicated. A review of four controlled clinical trials involving about one thousand individuals who were randomly allocated to either an antibiotic or a placebo showed that about 2 percent of placebo-treated subjects got Lyme disease in contrast to 0.2 percent in those who got the antibiotic.[3] This works out in theory to one case of Lyme disease prevented for every fifty tick bites that are treated. The likelihood of benefit goes up if the tick was visibly engorged with blood. Some may choose with justification to wait and see if they become symptomatic, especially if they are likely to get other tick bites during the spring or summer.

The recommended therapy for those choosing prophylaxis is simple: a single dose of doxycycline by mouth.[4] Although it is likely (but not yet proved) that the doxycycline would also reduce the risk of human granulocytic anaplasmosis (HGA) infection, this drug cannot be counted on to prevent babesiosis or a viral infection acquired from the tick. It appears that amoxicillin or penicillin similarly reduce the risk of illness after a tick bite, so either would be a suitable alternative to doxycycline for a child of 8 years or less, a pregnant or nursing woman, or someone who is allergic to tetracyclines. But the treatment durations with a penicillin antibiotic in the trials were 10 days, a period during which side effects are to be expected in some people. A penicillin-type drug would also not prevent HGA infection.

An experimental method not yet evaluated in humans is the topical application of an antibiotic cream containing a 4 percent concentration of the macrolide azithromycin. This preparation, but not a comparably dosed preparation of doxycycline, prevented *B. burgdorferi* in laboratory mice when it was applied to the bite site after tick removal.[5] There is not a commercial version of the azithromycin skin cream in the United States, but presumably it could be formulated by a compounding pharmacy from a doctor's prescription.

An Overview of Tick Repellents

Minimizing skin exposure with suitably long and sealed-off clothing will reduce tick bites, but parents, camp counselors, and job foremen, among others, know that such measures are not always readily or willingly adhered to. Tick activity in the northeastern and north-central United States is usually at its height during warmer months of the year. Many people opt to risk a tick bite rather than to be confined in uncomfortable long sleeves and pants. Whatever amount of skin is exposed, another protective measure is the use of repellents. Tick checks alone may not be enough.

A repellent is a substance that causes a tick, mosquito, or other arthropod pest to make an oriented movement away from the source—in less technical terms, to head in the other direction. Mosquitoes and ticks zero in on potential hosts, either from a distance or, once attached, using natural body odors or carbon dioxide as the attractants. The tick senses the volatile chemicals released by a potential host through specific receptors and nerve cells in organs that are analogous to the sensory apparatus in our noses. Ticks do not have antennae, which insects have for their sensory organs. Instead, these sense organs are located on forelegs and other appendages of the ticks. This sensory system informs the arthropod about where the next meal is located but also about places to avoid. Repellents stimulate the avoidance receptors, thus overriding the effects of attractants from the host. The repellent can have this effect in a gaseous phase, which would be desirable when mosquitoes or gnats are buzzing around your head, or on the skin surface, when the arthropod contacts the skin itself or clothing.

Most repellents in practice are topically applied, though they might also work through systemic distribution to the breath or skin after oral consumption. Some chemicals work as both an acaricide and a repellent, but most are one or the other. The ideal repellent should provide protection for at least eight hours, be effective against a variety of blood-feeding arthropods, have very low to nil toxicity, and be nongreasy, nonirritating, and odorless.[6] There are no repellents that currently meet all these criteria, though some come close.

There are some general observations to keep in mind when considering repellent choices.[7] The first is that most repellents are developed with mosquitoes and other flying insects in mind. Ticks are usually a secondary consideration. And when studies do include ticks, they less frequently test on an *Ixodes* species than on species of *Amblyomma, Dermacentor,* and *Rhipicephalus,* for which there is a larger repellent market in the world. These species distinctions may not be apparent in advertisements or the product's big print label, but it should be there in fine print on the container or at the company website. In any case, what we ask of repellents against ticks is not the same as the goals for repellents against flying insects. An *I. scapularis* nymph perched on a leaf may not be dissuaded by a repellent from grabbing onto your sock or bit of skin in that instant you walk by. It would be enough if the repellent repulsed that tick on your pants from reaching exposed skin or repulsed it from settling in to bite if it makes it there. Compare that with our expectations for a repellent against mosquitoes: we don't want mosquitoes to land at all.

It follows then that the methods for testing a repellent in the laboratory or the field are different for ticks than for mosquitoes. The methods vary in type from those limited to a test tube or petri dish to those entirely in the field. In one field test of tick repellents, people wear over-the-calf white socks, one of which has the repellent and the other just the solution used for dissolving the repellent. The testers then tramp through fields or other areas where ticks abound, and afterward they count the ticks on each leg that have ascended off the sock onto the lower leg. Somewhere between these types of tests are studies that are done under laboratory conditions with human or dog subjects. The EPA specifies the following type of human subject test of tick repellents if the repellent is to be registered for application to human skin*: a tick is released on untreated skin of the lower part of an arm held vertically, and whether a tick within three minutes crawls upward into a treated zone of skin for at least 3 centimeters is recorded.

This type of testing is typical of what is expected for approval as a repellent, and there are numerous research papers that use something like this method as the standard for evaluating candidate repellents. These laboratory tests have some value for predicting repellent success, to be sure, but just how well do they correlate with performance in typical outdoors situations during recreation, work, or around the home? There are few controlled research studies of tick repellents as they would be used in real life. In one of the better ones, 111 volunteers had 112 bites from *I. ricinus* when they were not wearing a repellent and 42 bites during an equivalently long period when a recommended tick repellent was applied in prescribed way.[8] These volunteers spent at least two hours outdoors in a high tick-density habitat in Sweden over the month-long study. That was a significant reduction in bites with the repellent, but this method of protection was by no means foolproof. The risk of disease transmission was lowered by more than half, but it still existed.

Finally, the formulation of the repellent—that is, the medium in which it is applied—can be as important as the performance of the repellent substance itself. A repellent may perform terrifically in tests in the laboratory with petri dishes or glass tubes, but if it too quickly evaporates when applied to human skin, or it is too greasy or has an unpleasant smell, it won't be adopted well. These cosmetic aspects are important, as the manufacturers recognize. Some repellents may be comparatively well tolerated on skin itself, but they are irritating if some gets in the eyes or are damaging to certain plastics or fabrics.

*U.S. Environmental Protection Agency OPPTS 810.3700: Insect Repellents to be Applied to Human Skin.

Specific Repellents

My 1996 book featured two chemicals for personal protection: DEET (*N,N*-diethyl-*m*-toluamide) and the pyrethroid permethrin. Permethrin is listed by the EPA as both a repellent and an insecticide, but it mainly succeeds in protecting people from tick and insect bites as an insecticide. This is an important distinction. Just as we would not think of putting an insecticide, like carbaryl, on our skin, so we should not use permethrin or another pyrethroid on skin. It is meant for and should be restricted to application on clothing fabrics for personal use. Permethrin formulations are available for consumers to treat their pants, socks, shirts, hats, and boots. There are also clothes products that already have the permethrin stably impregnated into the fabric, which retains potency over more launderings than clothing treated by consumers. (DEET can also be applied to clothing, except to certain synthetic fabrics, but it is has been supplanted by pyrethroids for this purpose.)

Joining DEET now on a list of repellents recommended by the CDC for protection against mosquitoes are three other compounds: picaridin, IR3535, and *p*-methane-3,8-diol (PMD). DEET, first marketed in 1957, remains the most widely used insect and tick repellent, with about 200 million users a year, but it has competition now from the three other repellents.[9] DEET is applied to the skin in concentrations of 10 to 100 percent, but only those of 20 percent or higher give significant protection for more than a few hours. The duration of the repellent's effect depends on the concentration but also varies with the formulation and on the conditions for evaporation, perspiration, and water exposure. Formulations in which the substance is contained in microscopic capsules ("microencapuslated") allow for sustained release of the active ingredient slowly over time. A microencapsulated formulation with 34 percent DEET repulses arthropods about as effectively and for as long as preparations containing up to 100 percent DEET. The microencapsulated form of DEET is somewhat less sticky and smelly than older formulations. A concentration of 20 percent DEET remains the benchmark for comparison in laboratory tests of repellents in development.[10]

DEET is generally safer than insecticides, such as permethrin, when it is confined to the skin, but rarely it can be toxic if ingested by mouth or absorbed through open cuts or the eyes. DEET is also absorbed into the blood through intact skin and can be found in the urine for days after a skin application. DEET can damage plastics and should not be smeared on plastic lenses, goggles, or watch crystals. The risk of adverse effects of DEET can be minimized by (1) avoiding concentrations higher than 50 percent, especially if it will be used for several days in a row; (2) steering clear of lips, eyes, open wounds, and inflamed skin; (3) not inhaling while spraying; (4) keeping DEET away from skin under clothing; (5) avoiding combinations of DEET with a sunscreen; (6) washing affected skin or bathing after returning home; and (7) using a lower concentration and avoiding the hands or near the eyes or mouths of young children. If the main concern is ticks instead of mosquitoes, then DEET or other repellent need not be applied to face or hands, thus minimizing the chance of inadvertent ingestion.

Picaridin, also known as KBR3023 and icaridin, is a derivative of the chemical piperidine. It was introduced in the United States as a repellent in 2005. Picaridin has the advantages of being odorless and nongreasy, and it does not damage plastics or fabrics. Concentrations range from 20 to 30 percent. There is less experience with picaridin than with DEET as a repellent in the United States than in Europe, but it appears to be as or nearly as effective as the older repellent.

IR3535 (ethyl butylacetylaminopropionate, or EBAAP) is almost like a natural product because it is derived from the amino acid beta-alanine. It is classified as a "biochemical" substance by the EPA. Although it is popular around the world and is found in more than 150 consumer products in concentrations of 10 to 35 percent, IR3535 has been slow to catch on in the United States, where it was not registered until 1999. It has no odor and dries on the skin without a residual greasy feel. According to the EPA's fact sheet, "IR3535 has been used in Europe for 20 years with no substantial adverse effects."[11] In limited studies to date, IR3535 was more effective than DEET against *Ixodes* species.[12] Two sources in North America are Avon's Skin-So-Soft Bug Guard brand and Coleman's SkinSmart brand.

PMD is a synthetic form of a chemical extracted from the lemon eucalyptus tree. The nonsynthetic botanical form of this chemical is registered with the EPA as "oil of lemon eucalyptus" and in the United Kingdom as Citridiol. It is generally used in concentrations of 20 to 60 percent for protection against insects and ticks. PMD is not irritating for the skin, but contact with the eyes should be avoided. Some people dislike the odor of some preparations. Products with oil of lemon eucalyptus should not be used on children under the age of 3 years.

Many other repellents based on natural products are available or under development. Most of these botanicals have low effectiveness to begin with or last for only a short time. One that is more promising and that is available in the United States is 2-undecanone (BioUD), which is derived from the wild tomato, *Lycopersicon hirsutum*. There has been much less testing of this product against ticks, but in one laboratory study, it was as effective as DEET for the Lone Star tick and the American dog tick.[13]

Pet Precautions

Pets should also be examined, especially around their eyes and ears, a favorite spot for ticks on furry animals. Some people use a floor mirror to examine dogs' undersides. Brushing or combing the pet's fur may reveal loose or attached ticks. Unattached ticks can be picked off with the fingers or tweezers. Attached ticks can be removed by the method used for people.

There are a variety of products for preventing or treating tick infestation on pets. Flea-and-tick collars seem to be a simple answer to the problem of preventing tick bites, but they are more useful for preventing fleas than ticks. Their success is greater for smaller animals than larger ones. Collars containing the insecticide carbaryl were discontinued because of safety concerns for pets and for people who inadvisably wore the collars themselves around their necks or limbs. The active ingredients in the four types of collars available in the United States are (1) imidacloprid and flumethrin for dogs; (2) amitraz for dogs; (3) deltamethrin for dogs; and

(4) the carbamate insecticide propoxur and the insect growth regulator [S]-methoprene for cats. (Pyrethroid and amitraz-based products should not be used for cats.)

An alternative to collars for dogs are spot-on insecticides, which are applied at about monthly intervals to one spot on the dog, usually the nape of the neck, and which after absorption are then dispersed through the blood to the skin. These insecticides are usually combinations of compounds. Some of the components of the formulation, such as imidacloprid and [S]-methoprene, are irrelevant for tick bite prevention; they are there for the fleas. But at least one other member of the combination, like fipronil or a pyrethroid, would be expected to be effective against ticks. Currently available products are the following: (1) fipronil and [S]-methoprene; (2) fipronil, [S]-methoprene, and cyphenothrin; (3) fipronil, [S]-methoprene, and amitraz; (4) fipronil and cyphenothrin; (5) imidacloprid and permethrin; and (6) selamectin. Selamectin can be used for both dogs and cats, but in one brand name formulation, it is only approved for prevention of American dog tick bites. There are imidacloprid plus pyriproxyfen spot-on flea treatments available for cats, but these would not work against ticks.

A medication that is taken orally in a flavored chew by the dog once a month is afoxolaner (NexGard),[14] which was shown to be effective against the American dog tick, Lone Star tick, *I. scapularis*, and *I. ricinus*.[15] Other oral medicines, like the natural insecticide spinosad, are available for dogs and cats for prevention of flea infestations. Spinosad is not yet approved for tick control, though it appears to have activity against ticks in dogs.[16]

Residential Landscape Management

The risk of infection around a residence can be partially mitigated by eliminating places where disease-carrying mosquitoes can breed, such as trays under pots for plants, used tires, discarded bottles, and some flowers that gather rainfall. These steps would be irrelevant for tick control, but there are analogously simple measures that households can take to reduce tick encounters on or near the property. These measures can also be carried out on a neighborhood or community-wide scale and on the grounds of workplaces, schools, and parks. Such domestic modifications may have a greater impact in the northeastern United States, where a greater proportion of the exposures are around residences, than in the north-central states, where exposures in recreational areas make a larger contribution.[17]

One can imagine a tick creeping up your pant leg or embedded in the ear of a dog, but in truth, almost all of a tick's life is spent off a host. Some types of ticks live in the nests or burrows of their preferred hosts, like chipmunks or ground squirrels. Such ticks are more dependent on the fates of the animals living in the burrow or tree snag than on outside environmental conditions. In contrast, questing ticks like *I. scapularis* and *A. americanum*—that is, those that put themselves out there on grass blades and leaves—are more at the mercy of the elements.[18] Deer ticks prefer temperate forests to jungles and deserts, an indication of the importance of humidity (neither too much nor too little), temperature, and shielding from direct sunlight. This is important for each stage of the life cycle. Even a change in relative humidity from 95 percent to 65 percent is enough to significantly affect the survival of larvae and nymphs of *I. scapularis*. Alterations of these microenvironments can be achieved through landscape management of habitats, such as by removing leaf litter, under which the moisture levels are higher than the air above. Keeping the lawn well-manicured reduces humidity at ground level and lets in more sunlight. An important zone for attention is the interface between a woodland and a lawn or grass area. Another general principle is to impair the travel of ticks from a more protected forested habitat onto a grass area where people gather.

Here are some possible modifications that follow these principles:[19] (1) use wood chips or plant shade-tolerant grass under shade trees; (2) trim trees and brush to open up wooded areas in and around areas of human activity, allowing sunlight to penetrate; (3) avoid landscape plantings that attract deer; (4) create a 3-foot (1-meter) or wider border of wood chips, mulch, stone, or gravel between turf and woods; (5) clear tall grasses and brush around residences and at the edges of lawns; and (6) remove old furniture, trash, or other places where ticks may hide from the yard.

A burn may substantially reduce the number of ticks around a residence, but the effect is temporary and would need to be done regularly for sustained reduction. Since these burns are done in or around densely populated residential areas, there may be little tolerance for this approach over time. Another downside of controlled burns of the landscape is that nontargeted animals, such as mammals and reptiles that nest near the ground and near houses, may be adversely affected.

Human Vaccines against Lyme Disease

When exposure to an infectious disease is unavoidable and when effective treatment is either nonexistent or partial, we look to a cornerstone of preventive medicine: a vaccine. Malaria again serves as a starting point for the topic. This parasite infection can be prevented by measures aimed at the vector, but now many of its mosquito vectors are resistant to the insecticides that once kept their numbers in check. The illness of malaria has been successfully treated with medicines, such as quinine and chloroquine, but in an arms race with humans, the parasites are becoming more resistant to treatments, even to those recently developed. As the standbys of prevention and treatment lose effectiveness, there is keener interest in developing a malaria vaccine.

The case for a human Lyme disease vaccine is less compelling; the *Ixodes* ticks are not yet resistant to acaricides, and the bacterium remains susceptible to the same antibiotics. But for many people at high risk by virtue of their residences, recreational activities, or occupations, *B. burgdorferi* infection is difficult to avoid. On this basis, as well as the morbidity of unsuspected or undiagnosed infection, there has been commercial and governmental interest in developing a human vaccine against Lyme disease for over twenty years. But, as for malaria, there is no available Lyme disease vaccine at present. Although the aim for both vaccines is to prevent an infection transmitted by an arthropod, the histories of vaccine development for each of these vector-borne infections could not be more divergent.

Malaria investigators recognized early on that development of a malaria vaccine was going to be challenging. But with some fits and starts, and after decades of research, progress toward this globally important goal is being made. In contrast, a promising candidate vaccine against *B. burgdorferi* practically fell into developers' laps. One of the first proteins of *B. burgdorferi* ever characterized, the outer surface protein A (OspA),[20]* was shown in experiments carried out within ten years of the agent's discovery to provide protection of laboratory mice against infection.[21] This was before it was even known what OspA's function in the cell is. But, it wasn't long before the rush was on between two large pharmaceutical companies to turn the OspA protein into a commercial vaccine for humans against Lyme disease.[22] Two controlled clinical trials involving thousands of subjects given the vaccine or a placebo injection were carried out in the northeastern United States, practically simultaneously in the mid-1990s. The results of the trials were similarly encouraging, at least by the criteria set for an effective malaria vaccine. But only one of the companies, GlaxoSmithKline, decided to pursue FDA approval for a vaccine against Lyme disease. Approval was gained in 1998, and the commercial vaccine, named LYMErix, was available across the United States and Canada until February 2002, when it was withdrawn from the market. The manufacturer reported that there was not enough demand to justify the continued manufacture and marketing of the vaccine.

The whole story of the rise and fall of the human Lyme disease vaccine would occupy a full chapter itself, if not a whole book. There were cumulative wounds for the vaccine during its short life, but the most grievous may have been the failure of the vaccine to gain a full recommendation by the CDC's Advisory Committee on Immunization Practices (ACIP).[23] Because LYMErix wasn't recommended by the ACIP for routine immunization, it wasn't covered by the National Vaccine Injury Compensation Program, a kind of pooled no-fault insurance to offset

---

*I am a listed inventor on U.S. and European patents on the use of OspA as a vaccine and on this basis received royalties from companies that licensed the patent for human and dog vaccines. The key patents expired in 2014.

settlements from lawsuits. Without the protection that many other vaccines have, the LYMErix vaccine was open to a full assault of personal injury litigation. Here is what I wrote in my 1996 book, while the vaccine was in the FDA application process: "Inasmuch as Lyme disease is rarely fatal, the public's acceptance of side effects or actual illness from a vaccine would be expected to be low. What with the amount of litigation against vaccines for proven killers such as whooping cough and diphtheria, how great would be the bottom line for a vaccine against a more benign infection?"

Several years into the twenty-first century, there is a resurgence of interest in a human vaccine.[24,25] Memories of the trainwreck of LYMErix recede. We recognize again that, short of moving out of the area or remaining indoors, *B. burgdorferi* infection is sometimes inescapable, no matter how conscientious one is about personal and household protection. In my view, we as a society missed the opportunity for individuals at high risk of infection, through residential, occupational, or recreational exposures, to have the option of a vaccine. There may be a better chance for a Lyme disease vaccine in Europe, where a seasonal vaccine against tick-borne encephalitis virus is widely accepted (chapter 11). Addition of another component to a vaccine formulation to guard against Lyme disease as well as TBE would not be a stretch. But I am not optimistic about the near-term resurrection of either the old Lyme disease vaccine or a new one in the United States. I agree with Gregory Poland of the Mayo Clinic that "future candidate Lyme disease vaccines are unlikely to be developed, tested, and used within the United States in the near future, thus leaving at-risk populations unprotected."[26] (This includes promising antitick vaccines that are directed at *Ixodes* tick proteins and could work by blocking transmission of not only *B. burgdorferi* but other deer tick pathogens.)

Vaccines for Dogs

Humans can't get a Lyme disease vaccine, but dogs can. No fewer than three commercial products are available by prescription from your veterinarian. Two vaccines are "bacterins," a common form of vaccine for pets and domestic animals. Bacterins are usually whole bacterial cells that have been recovered from a culture and then inactivated by heating or some other process that kills the bacteria. The vaccine preparations often contain culture medium constituents as well as the cells. In the case of cultures of *B. burgdorferi*, this would be cow albumin from the blood and rabbit serum, among many substances. Bacterins also comprise hundreds of proteins of the bacteria. Importantly for a *B. burgdorferi* vaccine, one of these is OspA, the basis of the only approved human vaccine.

The two bacterin vaccines are similar in containing two different strains of *B. burgdorferi* in a single preparation. The initial vaccination for both is two doses injected under the skin two to four weeks apart. Annual revaccination with a single dose is recommended. One of the bacterin vaccines is called LymeVax, and it was the first Lyme disease vaccine on the market when it was introduced in 1990 by Fort Dodge Laboratories. That company was taken over by the animal health division of Pfizer, the multinational pharmaceutical company. Subsequently, Pfizer spun off its animal health subsidiary as a separate company called Zoetis, which is now the manufacturer of record for LymeVax. According to the company website (in mid-2014), 31 million doses of the vaccine have been given to dogs to that date. The vaccine was approved by the U.S. Department of Agriculture (USDA) for dogs 9 weeks or older. The second bacterin vaccine is called NOBIVAC Lyme and is a product of Merck Animal Health. It was originally sold under the name of Galaxy Lyme, when it was produced by Schering-Plough Animal Health, before the merger of that company with Merck in 2009. This vaccine was approved for dogs 8 weeks or older. (In Europe there is a dog vaccine, Merilym 3 from Merial Limited, which is a bacterin comprising the three different species of the cause of Lyme disease on that continent: *B. afzelii, B. garinii,* and *B. burgdorferi.*)

The third dog vaccine currently available in the United States is RECOMBITEK Lyme, which is produced by Merial Limited. As the brand name suggests, this vaccine is based on a purified recombinant protein, in this case OspA. The vaccine is very similar to the LYMErix OspA-based vaccine that was approved by the FDA for human use. The difference is that RECOMBITEK has no added "adjuvant," which is a substance, such as an aluminum salt, commonly added to human and veterinary vaccines to boost the immune response. The dog OspA vaccine in its unadjuvanted feature was like the second human vaccine that underwent clinical trials for efficacy and safety but never went forward for approval.[27] In a blinded, controlled trial, 10- to 12-week-old mixed-breed dogs were vaccinated with OspA or a negative control substance and then experimentally challenged three weeks later with infected *I. scapularis* ticks. The twenty OspA-vaccinated dogs had no evidence of clinical disease, seropositive antibody responses, or the presence of the spirochete by culture or PCR. Of the ten control dogs, only two (20 percent) became discernibly ill, but all ten had evidence of infection by antibody test and by culture or PCR.[28] On this basis, as well as other data provided by the company, the vaccine was approved by the USDA for dogs 9 weeks or older. Two initial doses given two to three weeks apart are followed by annual revaccinations thereafter.

The bacterin vaccines underwent similar in-house studies in which dogs were vaccinated and experimentally infected. These experiments under controlled conditions, together with data on the safety and toxicity of the vaccine for dogs, were sufficient for USDA approval. The bar for approval of a vaccine for domestic animals by the USDA is generally lower than for approval of a human vaccine by the FDA. The Lyme disease vaccines for dogs did not have to go through the equivalent of the large clinical trials of the human vaccine, which calls for assessment of naturally acquired infections under day-to-day, real-life conditions and extensive monitoring for adverse effects in the subjects during and after the study.

There has been one published study of a vaccine, the bacterin LymeVax, in dogs in a high-risk Lyme disease community.[29] It was not a controlled trial; the dogs were not randomly assigned to vaccine or placebo, as a human trial would have specified. And so it was not blinded; the dog owner participants and the investigators knew what group the dogs were in. About 2,000 dogs who received 4,000 doses of the vaccine over the 20-month period were compared with about 4,500 dogs that were not vaccinated. The same proportion of both groups of dogs had been infected with *B. burgdorferi* before the study started, as determined by an antibody test. About 2 percent of the dogs that were vaccinated had minor reactions (such as facial swelling, hives, itchiness, or swelling at the vaccination site) that resolved within 72 hours of injection. By the end of the study period, 5 percent of the unvaccinated dogs had a diagnosis of symptomatic Lyme disease compared with a 1 percent incidence in vaccinated dogs. This can be expressed as an efficacy rate for the vaccine of about 80 percent, which is close to what was found in the human clinical trials and which was sufficient to gain FDA approval for a human vaccine.[30]

Since *B. burgdorferi* infection is not transmitted from dog to dog, the rationale for the vaccine cannot be contagion containment, such as in a kennel or animal rescue center. Immunization of one dog will not prevent Lyme disease in another dog. So a vaccine for dogs is optional. All three vaccines seem to work moderately well at preventing infection and, as dog vaccines go, without an intolerable amount of adverse effects. (Although there are not published studies on the postmarketing experience, all three vaccines have been on the market for several years, with millions of doses of each dispensed. If there were significant adverse effects, even if anecdotally reported, they would likely be apparent by now.)

That said, are millions of vaccine doses a year justified? Most experts who have reviewed the data and current practices think not. Of the forty-five members of an ad hoc panel of the American College of Veterinary Internal Medicine that was convened in 2005, fifteen years after the introduction of a dog vaccine, thirty-nine did not recommend the vaccine, two recommended it, and four used it rarely.[31] Here are some of the points of the critics in this and other articles on vaccination of dogs:

1. The experimental infection studies showed that most of the symptomatic illness, notably the lameness from arthritis, occurred in young dogs, yet the vaccine is only indicated for dogs 8 weeks or older.
2. In areas of heavy Lyme disease risk, more than 75 percent of dogs may be exposed to infected ticks, but only 5 percent of those exposed develop an illness compatible with *B. burgdorferi* infection.

3. Perhaps the best argument for vaccination is to prevent the serious kidney disease associated with *B. burgdorferi* infection in dogs, which can be fatal. But this complication is uncommon, and its mechanism is not well understood. If it is caused by a type of immune response gone awry, leading to kidney damage, then it is conceivable that vaccination would make matters worse.

On the other hand, experts on consensus and guideline panels are often specialists with referral practices who are not on the front lines of day-to-day practice. Primary care veterinarians seem more likely to administer the vaccine. Is there a middle ground? Certainly responsible veterinarians and public health experts would agree that the vaccine is not indicated for dogs living in areas where *B. burgdorferi* does not occur. Most would agree that immunizing your dog against *B. burgdorferi* will not protect the people in the household from infection from a tick carried in by the dog.

Whether or not asymptomatic dogs that are seropositive for exposure to *B. burgdorferi* should be treated with an antibiotic first before receiving the vaccine is more contentious. Some experts recommend treatment first. Others do not think that is necessary as long as the dog is asymptomatic. In any case, having Lyme disease before may not prevent another infection, since immunity is generally strain-specific. The vaccine may prevent a second infection in dogs who have had Lyme disease once before.

What about cats? There is no approved vaccine for cats. And horses? No to that, too, but some horses are getting the canine vaccine after the owner signs a waiver.

For a summary comment on dog vaccines, I can do no better than the 2012 advice of Dr. Michael Stone, a staff veterinarian and faculty member at Cummings School of Veterinary Medicine at Tufts University in Massachusetts: "I do not routinely encourage vaccination against Lyme, even though I practice in a region where the disease is endemic. That said, I do not hesitate to administer the vaccine when a pet owner requests it, because I believe it is safe. However, because I'm not convinced of its effectiveness in preventing Lyme disease, I would emphasize that tick control still remains important. There are other diseases that can be transmitted by tick bites—Lyme vaccine does not prevent those other infections. Tick control for all dogs in tick-infested locations is more important than vaccination. There are both collar and topical products that are safe and effective in deterring ticks."[32]

Prevention: A Personal View

In the 1970s, I was a foot soldier of sorts in the World Health Organization's global campaign to eradicate smallpox from Earth. I was stationed in India, working alongside local health officers and other volunteers.[33] At the time it would have been too costly and too slow to try to vaccinate everyone in that country against this fearful infection. Instead, efforts focused on vaccinating people in the village or neighborhood of a newly discovered case. The campaign targeted those who were most likely to be the next link in the chain of transmission. If one could interrupt transmission often enough, eventually the virus would have nowhere to go and would die out. In the end, this strategy worked; India was declared smallpox-free in 1977. But these actions were backed by the government, and occasionally police were needed to enforce a quarantine zone or a mandatory vaccination.

This was a community-wide action writ on a national scale. Smallpox was finally declared eradicated from the world in 1980. I do not think this would have happened if the eradication campaign had depended on individuals choosing whether to have an optional vaccine or not. A libertarian might claim that rational decision making and self-interest would have eventually led to smallpox's demise without a public health campaign backed by a democratically elected government. But I doubt it. Measles was next in the sights of public health officials for global eradication, but three decades later—in these more individualistic times—there are still measles outbreaks in the world, including in the United States.

The timeline for global smallpox eradication overlapped the first description of Lyme disease in North America and then discovery of its cause in 1981 (see the introduction). Since then, I have witnessed the growth of Lyme disease in frequency and extent on this continent. Many scientific advances in understanding the pathogen, its ecology, and how it causes disease have been made. But with a few isolated exceptions, these have not led to a significant reduction in risk of acquiring the infection. A moderately effective human vaccine has come and gone. Remaining prevention efforts are best characterized as a patchwork in their implementation and effects. There are protective measures that individuals and households are taking on their own to various degrees. Some communities are taking significant steps toward tick control, such as deer culls or area-wide acaricides, while others are not.

Make no mistake: neither *B. burgdorferi* nor any other Lyme disease species will be eradicated from Earth. Unlike smallpox, measles, or even malaria, human hosts are irrelevant to the success of these microorganisms in nature. The most that can be hoped for is a substantial, call it 50 to 90 percent, reduction of the risk. But as long as current conditions and attitudes persist, overall reductions will likely remain marginal. Perhaps the existing pockets of concerted and successful efforts will point the way to others. There might then be wider buy-in, combined with a willingness to compromise, to achieve community-wide prevention campaigns that make a real difference for people's health.

Trusted Internet Sites

Videos

American Lyme Disease Foundation, tick removal video
    www.aldf.com/videos.shtml

Proceedings of the Royal Society B: Biological Sciences, "How ticks get under your skin" video
    rspb.royalsocietypublishing.org/content/280/1773/20131758/suppl/DC1

United States

American Academy of Neurology (for Lyme Disease of the nervous system)
    http://www.neurology.org/content/69/1/91.full

American Academy of Pediatrics, Insect Repellents
    www.healthychildren.org/English/safety-prevention/at-play/pages/Insect-Repellents.aspx

American College of Physicians, Smart Medicine: Lyme Disease
    http://smartmedicine.acponline.org/content.aspx?gbosId=62

American Lyme Disease Foundation
    www.aldf.com

Centers for Disease Control and Prevention, Lyme Disease
    www.cdc.gov/lyme

Centers for Disease Control and Prevention (CDC), *Tickborne Diseases of the United States,* 2nd edition, 2014
    www.cdc.gov/lyme/resources/TickborneDiseases.pdf

Connecticut Agricultural Experiment Station (includes information on tick control)

www.ct.gov/caes

Environmental Protection Agency, Insect Repellents: Use and Effectiveness (use search tools for help in choosing a repellent)

http://cfpub.epa.gov/oppref/insect

Environmental Protection Agency, Pesticides

www.epa.gov/pesticides

Humane Society of the United States, Lyme Disease

www.humanesociety.org/animals/resources/facts/lyme_disease.html

Infectious Diseases Society of America, Lyme Disease (includes link to practice guidelines)

www.idsociety.org/Lyme

Institute of Medicine's 2011 report on Lyme disease and other tick-borne diseases

www.ncbi.nlm.nih.gov/books/NBK57020/pdf/TOC.pdf

National Institute of Allergy and Infectious Diseases, Lyme Disease

www.niaid.nih.gov/topics/lymeDisease

National Library of Medicine, Lyme Disease (also available in Spanish)

www.nlm.nih.gov/medlineplus/lymedisease.html

Quackwatch, Lyme Disease: Questionable Diagnosis and Treatment

www.quackwatch.org/01QuackeryRelatedTopics/lyme.html

University of Rhode Island Tick Encounter Resource Center

www.tickencounter.org/ticksmart/top_ticksmart_actions

Canada

Public Health Agency of Canada, Lyme Disease

www.phac-aspc.gc.ca/id-mi/lyme-eng.php (English);

www.phac-aspc.gc.ca/id-mi/lyme-fra.php (Français)

Europe

European Concerted Action on Lyme Borreliosis
www.eucalb.com

Public Health England, guidelines for health professionals on Lyme disease
https://www.gov.uk/government/collections/lyme-disease
-guidance-data-and-analysis

Public Health England, information on tick bite risks and prevention
www.hpa.org.uk/Publications/InfectiousDiseases/EmergingInfections
/1304Ticksandyourhealth

Notes

Introduction. Three Case Histories and a Parable

1. K. Marcus, "Verhandlungen der dermatologischen Gesellschaft zu Stockholm," *Archives of Dermatological Research* 101 (1910): 403–406, link. springer.com/article/10.1007%2FBF01832773?LI=true.

2. G. J. Dammin, "Erythema migrans: A chronicle," *Reviews of Infectious Diseases* 11 (1989): 142–151, doi:10.2307/4454758.

3. K. Weber, "Remarks on the infectious disease caused by *Borrelia burgdorferi*," *Zentralblatt für Bakteriologie, Mikrobiologie, und Hygiene, Series A* 263 (1986): 206–208.

4. W. Burgdorfer et al., "Lyme disease—a tick-borne spirochetosis?" *Science* 216 (1982): 1317–1319, doi:10.1126/science.7043737.

5. B. Rensberger, "A new type of arthritis found in Lyme," *New York Times*, July 18, 1976.

6. A. C. Steere et al., "An epidemic of oligoarticular arthritis in children and adults in three Connecticut communities," *Arthritis and Rheumatism* 20 (1977): 7–17, doi:10.1002/art.1780200102.

7. A. C. Steere et al., "Antibiotic therapy in Lyme disease," *Annals of Internal Medicine* 93 (1980): 1–8, doi:10.7326/0003-4819-93-1-1.

8. J. A. Edlow, *Bull's-Eye: Unraveling the Medical Mystery of Lyme Disease* (New Haven, CT: Yale University Press, 2004).

9. D. Hurley and M. Santora, "Unproved Lyme disease tests prompt warnings," *New York Times*, August 23, 2005.

10. T. Burge, "Individualism and the mental," in *Midwest Studies in Philosophy*, vol. 4, ed. P. French, T. Uehling, and H. Wettstein (Minneapolis:

University of Minnesota Press, 1979), 73–121, doi:10.1111/j.1475-4975.1979
.tb00374.x.

11. J. Beall, Scholarly open access: Critical analysis of scholarly open-
access publishing, accessed October 11, 2014, http://scholarlyoa.com.

12. A. G. Barbour and D. Fish, "The biological and social phenomenon
of Lyme disease," *Science* 260 (1993): 1610–1616, doi:10.1126/science.
8503006.

Chapter 1. Early Infection and the Immune Response

1. W. E. Mast and W. M. Burrows, "Erythema chronicum migrans in
the United States," *Journal of the American Medical Association* 236 (1976):
859–860, doi:10.1001/jama.1976.03270080041031.

2. D. N. Naversen and L. W. Gardner, "Erythema chronicum migrans
in America," *Archives of Dermatology* 114 (1978): 253–254, doi:10.1001/archderm
.1978.01640140071018.

3. J. L. Benach et al., "Spirochetes isolated from the blood of two
patients with Lyme disease," *New England Journal of Medicine* 308 (1983):
740–742, doi:10.1056/nejm198303313081302.

4. E. D. Shapiro et al., "A controlled trial of antimicrobial prophylaxis
for Lyme disease after deer-tick bites," *New England Journal of Medicine* 327
(1992): 1769–1773, doi:10.1056/nejm199212173272501.

5. R. P. Smith et al., "Clinical characteristics and treatment outcome of
early Lyme disease in patients with microbiologically confirmed erythema
migrans," *Annals of Internal Medicine* 136 (2002): 421–428, doi:10.7326/0003
-4819-136-6-200203190-00005.

6. D. Liveris et al., "Improving the yield of blood cultures from patients
with early Lyme disease," *Journal of Clinical Microbiology* 49 (2011): 2166–
2168, doi:10.1128/jcm.00350-11.

7. A. C. Steere et al., "Vaccination against Lyme disease with recombinant *Borrelia burgdorferi* outer-surface lipoprotein A with adjuvant: Lyme
Disease Vaccine Study Group," *New England Journal of Medicine* 339 (1998):
209–215, doi:10.1056/nejm199807233390401.

8. L. E. Markowitz et al., "Lyme disease during pregnancy," *Journal of
the American Medical Association* 255 (1986): 3394–3396, doi:10.1001/jama.
1986.03370240064038.

9. V. Maraspin et al., "Erythema migrans in pregnancy," *Wiener
klinische Wochenschrift* 111 (1999): 933–940.

10. I. Mylonas, "Borreliosis during pregnancy: A risk for the unborn child?" *Vector Borne and Zoonotic Diseases* 11 (2011): 891–898, doi:10.1089/vbz.2010.0102.

11. N. van Burgel et al., "Severe course of Lyme neuroborreliosis in an HIV-1 positive patient: Case report and review of the literature," *BioMed Central Neurology* 10 (2010): 117, doi:10.1186/1471-2377-10-117.

Chapter 2. Late Infection and Its Complications

1. A. R. Pachner, P. Duray, and A. C. Steere, "Central nervous system manifestations of Lyme disease," *Archives of Neurology* 46 (1989): 790–795, doi:10.1001/archneur.1989.00520430086023.

2. J. P. Lawson and A. C. Steere, "Lyme arthritis: Radiologic findings," *Radiology* 154 (1985): 37–43, doi:10.1148/radiology.154.1.3964949.

3. H. F. McAlister et al., "Lyme carditis: An important cause of reversible heart block," *Annals of Internal Medicine* 110 (1989): 339–345, doi:10.7326/0003-4819-110-5-339.

4. S. I. Brummit et al., "Molecular characterization of *Borrelia burgdorferi* from a case of autochthonous Lyme arthritis in California," *Emerging Infectious Diseases*, 20 (2014): 2168–2170, doi:10.3201/eid2012.140655.

5. J. C. Miller et al., "Gene expression profiling provides insights into the pathways involved in inflammatory arthritis development: Murine model of Lyme disease," *Experimental Molecular Pathology* 85 (2008): 20–27, doi:10.1016/j.yexmp.2008.03.004.

6. A. C. Steere, E. E. Drouin, and L. J. Glickstein, "Relationship between immunity to *Borrelia burgdorferi* outer-surface protein A (OspA) and Lyme arthritis," *Clinical Infectious Diseases* 52, supplement 3 (2011): s259–s265, doi:10.1093/cid/ciq117.

7. Centers for Disease Control and Prevention, "Three sudden cardiac deaths associated with Lyme carditis—United States, November 2012–July 2013," *Morbidity and Mortality Weekly Report* 62 (2013): 993–996, www.cdc.gov/mmwr/preview/mmwrhtml/mm6249a1.htm.

8. R. L. Lesser, "Ocular manifestations of Lyme disease," *American Journal of Medicine* 98, supplement 4A (1995): 60S–62S, doi:10.1016/S0002-9343(99)80045-X.

9. Z. L. Berrada and S. R. Telford III, "Burden of tick-borne infections on American companion animals," *Topics in Companion Animal Medicine* 24 (2009): 175–181, doi:10.1053/j.tcam.2009.06.005.

Chapter 3. The Pathogen, Its Vector, and Its Reservoirs

1. A. G. Barbour and S. F. Hayes, "Biology of *Borrelia* species," *Microbiological Reviews* 50 (1986): 381–400, www.ncbi.nlm.nih.gov/pmc/articles /PMC373079/.

2. Ø. Brorson and S. H. Brorson, "Transformation of cystic forms of *Borrelia burgdorferi* to normal, mobile spirochetes," *Infection* 25 (1997): 240–246, link.springer.com/article/10.1007/BF01713153.

3. Oxford English Dictionary, "species, n.," Oxford University Press, June 2014 update, www.oed.com.

4. D. S. Samuels and J. D. Radolf, *Borrelia: Molecular Biology, Host Interaction and Pathogenesis* (Norfolk, UK: Caister Academic Press, 2010).

5. A. C. Steere and S. E. Malawista, "Cases of Lyme disease in the United States: Locations correlated with distribution of *Ixodes dammini*," *Annals of Internal Medicine* 91 (1979): 730–733, doi:10.7326/0003-4819-91-5-730.

6. T. Say, J. L. Le Comte, and G. Ord, *The Complete Writings of Thomas Say on the Entomology of North America* (Philadelphia, PA: A. E. Foote, 1891).

7. R. S. Lane and G. B. Quistad, "Borreliacidal factor in the blood of the western fence lizard (*Sceloporus occidentalis*)," *Journal of Parasitology* 84 (1998): 29–34, doi:10.2307/3284524.

8. B. Olsén et al., "A Lyme borreliosis cycle in seabirds and *Ixodes uriae* ticks," *Nature* 362 (1993): 340–342, doi:10.1038/362340a0.

9. J. Bunikis et al., "*Borrelia burgdorferi* infection in a natural population of *Peromyscus leucopus* mice: A longitudinal study in an area where Lyme borreliosis is highly endemic," *Journal of Infectious Diseases* 189 (2004): 1515–1523, doi:10.1086/382594.

10. E. Baum, F. Hue, and A. G. Barbour, "Experimental infections of the reservoir species *Peromyscus leucopus* with diverse strains of *Borrelia burgdorferi*, a Lyme disease agent," *MBio* 3 (2012): e00434–e00412, doi:10.1128/mBio. 00434-12.

11. D. Brisson, D. E. Dykhuizen, and R. S. Ostfeld, "Conspicuous impacts of inconspicuous hosts on the Lyme disease epidemic," *Proceedings of the Royal Society B: Biological Sciences* 275 (2008): 227–235, doi:10.1098/rspb. 2007.1208.

12. R. P. Smith, Jr., et al., "Norway rats as reservoir hosts for Lyme disease spirochetes on Monhegan Island, Maine," *Journal of Infectious Diseases* 168 (1993): 687–691, doi:10.1093/infdis/168.3.687.

13. R. S. Lane et al., "Western gray squirrel (Rodentia: Sciuridae): A primary reservoir host of *Borrelia burgdorferi* in Californian oak woodlands?" *Journal of Medical Entomology* 42 (2005): 388–396, doi:10.1603/0022-2585(2005)042[0388:WGSRSA]2.0.CO;2.

Chapter 4. The Ecology of Lyme Disease

1. Anonymous, "Brought deer to Nantucket," *Inquirer and Mirror* (Nantucket Island, MA), June 10, 1922.

2. P. Belluck, "Tick-borne illnesses have Nantucket considering some deer-based solutions," *New York Times*, September 5, 2009.

3. J. Graziadei, "Nantucket deer herd larger than estimated," *Inquirer and Mirror* (Nantucket Island, MA), May 19, 2013.

4. A. Spielman, J. Piesman, and P. Etkind, "Epizootiology of human babesiosis," *Journal of the New York Entomological Society* 85 (1977): 214–216.

5. B. P. Flanagan, "Erythema chronicum migrans Afzelius in Americans," *Archives of Dermatology* 86 (1962): 410–411, doi:10.1001/archderm.1962.01590100024007.

6. W. E. Mast and W. M. Burrows, "Erythema chronicum migrans in the United States," *Journal of the American Medical Association* 236 (1976): 859–860, doi:10.1001/jama.1976.03270080041031.

7. A. G. Hoen et al., "Phylogeography of *Borrelia burgdorferi* in the eastern United States reflects multiple independent Lyme disease emergence events," *Proceedings of the National Academy of Sciences U.S.A.* 106 (2009): 15013–15018, doi:10.1073/pnas.0903810106.

8. F. Larrousse, A. G. King, and S. B. Wolbach, "The overwintering in Massachusetts of *Ixodiphagus caucurtei*," *Science* 67 (1928): 351–353, doi:10.1126/science.67.1735.351.

9. D. H. Persing et al., "Detection of *Borrelia burgdorferi* DNA in museum specimens of *Ixodes dammini* ticks," *Science* 249 (1990): 1420–1423, doi:10.1126/science.2402635.

10. K. M. Pepin et al., "Geographic variation in the relationship between human Lyme disease incidence and density of infected host-seeking *Ixodes scapularis* nymphs in the Eastern United States," *American Journal of Tropical Medicine and Hygiene* 86 (2012): 1062–1071, doi:10.4269/ajtmh.2012.11-0630.

11. A. G. Barbour and B. Travinsky, "Evolution and distribution of the OspC gene, a transferable serotype determinant of *Borrelia burgdorferi*," *MBio* 1 (2010): e00153-10, doi:10.1128/mBio.00153-10.

12. N. H. Ogden et al., "The emergence of Lyme disease in Canada," *Canadian Medical Association Journal* 180 (2009): 1221–1224, doi:10.1503/cmaj.080148.

13. B. Evengård and R. Sauerborn, "Climate change influences infectious diseases both in the Arctic and the tropics: Joining the dots," *Global Health Action* 2 (November 11, 2009), doi:10.3402/gha.v2i0.2106.

14. T. G. Jaenson and E. Lindgren, "The range of *Ixodes ricinus* and the risk of contracting Lyme borreliosis will increase northwards when the vegetation period becomes longer," *Ticks and Tick-Borne Diseases* 2 (2011): 44–49, doi:10.1016/j.ttbdis.2010.10.006.

15. C. Finch et al., "Integrated assessment of behavioral and environmental risk factors for Lyme disease infection on Block Island, Rhode Island," *PLoS One* 9 (2014): e84758, doi:10.1371/journal.pone.0084758.

16. Centers for Disease Control and Prevention, "CDC provides estimate of Americans diagnosed with Lyme disease each year," press release, August 19, 2013, www.cdc.gov/media/releases/2013/p0819-lyme-disease.html.

17. A. F. Hinckley et al., "Lyme disease testing by large commercial laboratories in the United States," *Clinical Infectious Diseases* 59 (2014): 676–681, doi:10.1093/cid/ciu397.

18. J. Tierney, "B. I. has high Lyme disease rates," *Block Island Times* (RI), September 25, 2010.

19. G. F. Eaton, "The prehistoric fauna of Block Island, as indicated by its ancient shell heaps," *American Journal of Science* 32 (1898): 137–159.

20. M. Tveskov, "Maritime settlement and subsistence along the southern New England coast: evidence from Block Island, Rhode Island," *North American Archaeologist* 18 (1997): 343–361, doi:10.2190/UHQG-34BU-W256-GLCR.

21. R. Salit, "Block Island culling deer herd with hunting-by-lottery, bait-and-shoot effort," *Providence Journal* (RI) January 24, 2014.

Chapter 5. Approach to Diagnosis

1. World Health Organization, "International Classification of Diseases (ICD)," accessed October 11, 2014, www.who.int/classifications/icd/en.

2. C. D. Tibbles and J. A. Edlow, "Does this patient have erythema migrans?" *Journal of the American Medical Association* 297 (2007): 2617–2627, doi:10.1001/jama.297.23.2617.

3. G. Stanek et al., "Lyme borreliosis," *Lancet* 379 (2012): 461–473, doi:10.1016/S0140-6736(11)60103-7.

4. G. P. Wormser, "Clinical practice: Early Lyme disease," *New England Journal of Medicine* 354 (2006): 2794–2801, doi:10.1056/NEJMcp061181.

Chapter 6. Laboratory Tests

1. G. Minot, discussion of F. W. Peabody, "The physician and the laboratory," *Boston Medical and Surgical Journal* 187 (1922): 324–327, doi:10.1056/NEJM192208311870905.

2. P. J. Krause et al., "*Borrelia miyamotoi* sensu lato seroreactivity and seroprevalence in the northeastern United States," *Emerging Infectious Disease* 20 (2014): 1183–1190, doi:10.3201/eid2007.131587.

3. A. C. Steere et al., "Prospective study of serologic tests for Lyme disease," *Clinical Infectious Diseases* 47 (2008): 188–195, doi:10.1086/589242.

4. K. L. Jones et al., "Strong IgG antibody responses to *Borrelia burgdorferi* glycolipids in patients with Lyme arthritis, a late manifestation of the infection," *Clinical Immunology* 132 (2009): 93–102, doi:10.1016/j.clim.2009.03.510.

5. A. G. Barbour et al., "Antibodies of patients with Lyme disease to components of the *Ixodes dammini* spirochete," *Journal of Clinical Investigation* 72 (1983): 504–515, doi:10.1172/JCI110998.

6. Association of State and Territorial Public Health Laboratory Directors and the Centers for Disease Control and Prevention, *Proceedings of the Second National Conference on Serologic Diagnosis of Lyme Disease*, Dearborn, Michigan, October 27–29, 1994, p. 108.

7. Centers for Disease Control and Prevention, "Recommendations for test performance and interpretation from the Second National Conference on Serologic Diagnosis of Lyme Disease," *Morbidity and Mortality Weekly Report* 44 (1995): 590–591, www.cdc.gov/mmwr/preview/mmwrhtml/00038469.htm.

8. J. A. Branda et al., "2-tiered antibody testing for early and late Lyme disease using only an immunoglobulin G blot with the addition of a VlsE band as the second-tier test," *Clinical Infectious Diseases* 50 (2010): 20–26, doi:10.1086/648674.

9. T. Cerar et al., "Diagnostic value of cytokines and chemokines in Lyme neuroborreliosis," *Clinical Vaccine and Immunology* 20 (2013): 1578–1584, doi:10.1128/cvi.00353-13.

10. J. L. Benach et al., "Spirochetes isolated from the blood of two patients with Lyme disease," *New England Journal of Medicine* 308 (1983): 740–742, doi:10.1056/nejm198303313081302.

11. A. C. Steere et al., "The spirochetal etiology of Lyme disease," *New England Journal of Medicine* 308 (1983): 733–740, doi:10.1056/nejm198303313081301.

12. D. Liveris et al., "Comparison of five diagnostic modalities for direct detection of *Borrelia burgdorferi* in patients with early Lyme disease," *Diagnostic Microbiology and Infectious Diseases* 73 (2012): 243–245, doi:10.1016/j.diagmicrobio.2012.03.026.

13. E. Sapi et al., "Improved culture conditions for the growth and detection of *Borrelia* from human serum," *International Journal of Medical Science* 10 (2013): 362–376, doi:10.7150/ijms.5698.

14. B. J. Johnson, M. A. Pilgard, and T. M. Russell, "Assessment of new culture method for detection of *Borrelia* species from serum of lyme disease patients," *Journal of Clinical Microbiology* 52 (2014): 721–724, doi:10.1128/jcm.01674-13.

15. A. C. Steere, "Diagnosis and treatment of Lyme arthritis," *Medical Clinics of North America* 81 (1997): 179–194.

Chapter 7. Putting Laboratory Testing in Its Place

1. W. Burgdorfer et al., "Lyme disease-a tick-borne spirochetosis?" *Science* 216 (1982): 1317–1319, doi:10.1126/science.7043737.

2. C. M. Fraser et al., "Genomic sequence of a Lyme disease spirochaete, *Borrelia burgdorferi*," *Nature* 390 (1997): 580–586, doi:10.1038/37551.

3. A. G. Barbour, "Immunochemical analysis of Lyme disease spirochetes," *Yale Journal of Biology and Medicine* 57 (1984): 581–586, www.ncbi.nlm.nih.gov/pmc/articles/PMC2590000/.

4. A. G. Barbour et al., "Antibodies of patients with Lyme disease to components of the *Ixodes dammini* spirochete," *Journal of Clinical Investigation* 72 (1983): 504–515, doi:10.1172/JCI110998.

5. F. Dressler et al., "Western blotting in the serodiagnosis of Lyme disease," *Journal of Infectious Diseases* 167 (1993): 392–400, doi:10.1093/infdis/167.2.392.

6. A. F. Hinckley et al., "Lyme disease testing by large commercial laboratories in the United States," *Clinical Infectious Diseases* 59 (2014): 676–681, doi:10.1093/cid/ciu397.

7. J. A. Branda et al., "Performance of United States serologic assays in the diagnosis of Lyme borreliosis acquired in Europe," *Clinical Infectious Diseases* 57 (2013): 333–340, doi:10.1093/cid/cit235.

8. G. P. Wormser et al., "Utility of serodiagnostics designed for use in the United States for detection of Lyme borreliosis acquired in Europe and vice versa," *Medical Microbiology and Immunology* 203 (2014): 65–71, doi:10.1007/s00430-013-0315-0.

9. M. S. Klempner et al., "Intralaboratory reliability of serologic and urine testing for Lyme disease," *American Journal of Medicine* 110 (2001): 217–219, doi:10.1016/S0002-9343(00)00701-4.

10. H. Huppertz et al., "Rational diagnostic strategies for Lyme borreliosis in children and adolescents: Recommendations by the Committee for Infectious Diseases and Vaccinations of the German Academy for Pediatrics and Adolescent Health," *European Journal of Pediatrics* 171 (2012): 1619–1624, doi:10.1007/s00431-012-1779-4.

Chapter 8. Antibiotics and Lyme Disease

1. C. A. Walsh, E. W. Mayer, and L. V. Baxi, "Lyme disease in pregnancy: Case report and review of the literature," *Obstetrical and Gynecological Survey* 62 (2007): 41–50, doi:10.1097/01.ogx.0000251024.43400.9a.

2. R. Patel et al., "Death from inappropriate therapy for Lyme disease," *Clinical Infectious Diseases* 31 (2000): 1107–1109, doi:10.1086/318138.

3. S. I. Brummit et al., "Molecular characterization of *Borrelia burgdorferi* from a case of autochthonous Lyme arthritis in California," *Emerging Infectious Diseases*, 20 (2014): 2168–2170, doi:10.3201/eid2012.140655.

4. R. C. Johnson et al., "Taxonomy of the Lyme disease spirochetes," *Yale Journal of Biology and Medicine* 57 (1984): 529–537, www.ncbi.nlm.nih.gov/pmc/articles/PMC2590029/.

5. E. Sapi et al., "Evaluation of in-vitro antibiotic susceptibility of different morphological forms of *Borrelia burgdorferi*," *Infection and Drug Resistance* 4 (2011): 97–113, doi:10.2147/idr.s19201.

6. G. P. Wormser and I. Schwartz, "Antibiotic treatment of animals infected with *Borrelia burgdorferi*," *Clinical Microbiological Reviews* 22 (2009): 387–395, doi:10.1128/cmr.00004-09.

7. R. J. Kazragis et al., "In vivo activities of ceftriaxone and vancomycin against *Borrelia* spp. in the mouse brain and other sites," *Antimicrobial Agents and Chemotherapy* 40 (1996): 2632–2636, www.ncbi.nlm.nih.gov /pmc/articles/PMC163589.

8. A. C. Steere et al., "Antibiotic therapy in Lyme disease," *Annals of Internal Medicine* 93 (1980): 1–8, doi:10.7326/0003-4819-93-1-1.

9. A. C. Steere et al., "Successful parenteral penicillin therapy of established Lyme arthritis," *New England Journal of Medicine* 312 (1985): 869–874, doi:10.1056/nejm198504043121401.

10. G. P. Wormser et al., "The clinical assessment, treatment, and prevention of lyme disease, human granulocytic anaplasmosis, and babesiosis: Clinical practice guidelines by the Infectious Diseases Society of America," *Clinical Infectious Diseases* 43 (2006): 1089–1134, doi:10.1086/508667.

11. B. W. Berger et al., "Failure of *Borrelia burgdorferi* to survive in the skin of patients with antibiotic-treated Lyme disease," *Journal of the American Academy of Dermatology* 27 (1992): 34–37, doi:10.1016/0190-9622(92)70152-6.

12. U. Ljostad et al., "Oral doxycycline versus intravenous ceftriaxone for European Lyme neuroborreliosis: A multicentre, non-inferiority, double-blind, randomised trial," *Lancet Neurology* 7 (2008): 690–695, doi:10.1016/ s1474-4422(08)70119-4.

13. S. W. Luger et al., "Comparison of cefuroxime axetil and doxycycline in treatment of patients with early Lyme disease associated with erythema migrans," *Antimicrobial Agents and Chemotherapy* 39 (1995): 661–667, doi:10.1128/AAC.39.3.661.

14. D. S. Samuels, K. E. Mach, and C. F. Garon, "Genetic transformation of the Lyme disease agent *Borrelia burgdorferi* with coumarin-resistant gyrB," *Journal of Bacteriology* 176 (1994): 6045–6049, jb.asm.org/content /176/19/6045.full.pdf+html.

Chapter 9. Putting Antibiotics to Use

1. J. P. Hanrahan et al., "Incidence and cumulative frequency of endemic Lyme disease in a community," *Journal of Infectious Diseases* 150 (1984): 489–496, doi:10.1093/infdis/150.4.489.

2. R. A. Kalish et al., "Evaluation of study patients with Lyme disease, 10-20-year follow-up," *Journal of Infectious Diseases* 183 (2001): 453–460, doi:10.1086/318082.

3. E. Zhioua et al., "Longitudinal study of Lyme borreliosis in a high risk population in Switzerland," *Parasite* 5 (1998): 383–386.

4. H. Kuiper et al., "One year follow-up study to assess the prevalence and incidence of Lyme borreliosis among Dutch forestry workers," *European Journal of Clinical Microbiology and Infectious Diseases* 12 (1993): 413–418.

5. Z. L. Berrada and S. R. Telford III, "Burden of tick-borne infections on American companion animals," *Topics in Companion Animal Medicine* 24 (2009): 175–181, doi:10.1053/j.tcam.2009.06.005.

6. M. P. Littman et al., "ACVIM small animal consensus statement on Lyme disease in dogs: Diagnosis, treatment, and prevention," *Journal of Veterinary Internal Medicine* 20 (2006): 422–434, doi:10.1111/j.1939-1676.2006.tb02880.x.

7. B. Gerber et al., "Follow-up of Bernese mountain dogs and other dogs with serologically diagnosed *Borrelia burgdorferi* infection: What happens to seropositive animals?" *BioMed Central Veterinary Research* 5 (2009): 18, doi:10.1186/1746-6148-5-18.

8. R. K. Straubinger et al., "Status of *Borrelia burgdorferi* infection after antibiotic treatment and the effects of corticosteroids: an experimental study," *Journal of Infectious Diseases* 181 (2000): 1069–1081, doi:10.1086/315340.

9. G. P. Wormser et al., "The clinical assessment, treatment, and prevention of lyme disease, human granulocytic anaplasmosis, and babesiosis: Clinical practice guidelines by the Infectious Diseases Society of America," *Clinical Infectious Diseases* 43 (2006): 1089–1134, doi:10.1086/508667.

10. P. M. Lantos et al., "Final report of the Lyme disease review panel of the Infectious Diseases Society of America," *Clinical Infectious Diseases* 51 (2010): 1–5, doi:10.1086/654809.

11. J. J. Halperin et al., "Practice parameter: Treatment of nervous system Lyme disease (an evidence-based review): Report of the Quality Standards Subcommittee of the American Academy of Neurology," *Neurology* 69 (2007): 91–102, doi:10.1212/01.wnl.0000265517.66976.28

12. ILADS Working Group, "Evidence-based guidelines for the management of Lyme disease," *Expert Reviews of Anti-infective Therapy* 2, supplement 1 (2004): S1–S13, www.ilads.org/lyme/treatment-guideline.php.

Chapter 10. After Antibiotic Therapy Ends

1. H. Dinerman and A. C. Steere, "Lyme disease associated with fibromyalgia," *Annals of Internal Medicine* 117 (1992): 281–285, doi:10.7326/0003-4819-117-4-281.

2. J. N. Aucott, L. A. Crowder, and K. B. Kortte, "Development of a foundation for a case definition of post-treatment Lyme disease syndrome," *International Journal of Infectious Disease* 17 (2013): e443–e449, doi:10.1016/j.ijid.2013.01.008.

3. E. G. Seltzer et al., "Long-term outcomes of persons with Lyme disease," *Journal of the American Medical Association* 283 (2000): 609–616, doi:10.1001/jama.283.5.609.

4. E. D. Shapiro et al., "Response to meta-analysis of Lyme borreliosis symptoms," *International Journal of Epidemiology* 34 (2005): 1437–1439, doi:10.1093/ije/dyi241.

5. J. N. Aucott et al., "Post-treatment Lyme disease syndrome symptomatology and the impact on life functioning: Is there something here?" *Quality of Life Research* 22 (2013): 75–84, doi:10.1007/s11136-012-0126-6.

6. I. Hickie et al., "Post-infective and chronic fatigue syndromes precipitated by viral and non-viral pathogens: Prospective cohort study," *British Medical Journal* 333 (2006): 575, doi:10.1136/bmj.38933.585764.AE.

7. S. Galbraith et al., "Peripheral blood gene expression in postinfective fatigue syndrome following from three different triggering infections," *Journal of Infectious Diseases* 204 (2011): 1632–1640, doi:10.1093/infdis/jir612.

8. A. G. Barbour, *Lyme Disease: The Cause, The Cure, The Controversy* (Baltimore, MD: Johns Hopkins University Press, 1996).

9. B. Fallon et al., "A randomized, placebo-controlled trial of repeated IV antibiotic therapy for Lyme encephalopathy," *Neurology* 70 (2008): 992–1003, doi:10.1212/01.WNL.0000284604.61160.2d.

10. R. F. Kaplan et al., "Cognitive function in post-treatment Lyme disease: Do additional antibiotics help?" *Neurology* 60 (2003): 1916–1922, doi:10.1212/01.WNL.0000068030.26992.25.

11. M. S. Klempner et al., "Two controlled trials of antibiotic treatment in patients with persistent symptoms and a history of Lyme disease," *New England Journal of Medicine* 345 (2001): 85–92, doi:10.1056/NEJM200107123450202.

12. L. B. Krupp et al., "Study and treatment of post Lyme disease (STOP-LD): A randomized double masked clinical trial," *Neurology* 60 (2003): 1923–1930, doi:10.1212/01.WNL.0000071227.23769.9E.

13. A. K. Delong et al., "Antibiotic retreatment of Lyme disease in patients with persistent symptoms: A biostatistical review of randomized, placebo-controlled, clinical trials," *Contemporary Clinical Trials* 33 (2012): 1132–1142, doi:10.1016/j.cct.2012.08.009.

14. M. S. Klempner et al., "Treatment trials for post–Lyme disease symptoms revisited," *American Journal of Medicine* 126 (2013): 665–669, doi:10.1016/j.amjmed.2013.02.014.

15. J. Sjowall et al., "Doxycycline-mediated effects on persistent symptoms and systemic cytokine responses post-neuroborreliosis: A randomized, prospective, cross-over study," *BioMed Central Infectious Diseases* 12 (2012): 186, doi:10.1186/1471-2334-12-186.

16. R. B. Stricker et al., "Benefit of intravenous antibiotic therapy in patients referred for treatment of neurologic Lyme disease," *International Journal of General Medicine* 4 (2011): 639–646, doi:10.2147/ijgm.s23829.

17. Y. Gu et al., "Non-antibacterial tetracycline formulations: Clinical applications in dentistry and medicine," *Journal of Oral Microbiology* 4 (2012), doi:10.3402/jom.v4i0.19227.

18. D. Romero et al., "Antibiotics as signal molecules," *Chemical Reviews* 111 (2011): 5492–5505, doi:10.1021/cr2000509.

19. E. J. Cameron et al., "Long-term macrolide treatment of chronic inflammatory airway diseases: Risks, benefits and future developments," *Clinical and Experimental Allergy* 42 (2012): 1302–1312, doi:10.1111/j.1365-2222.2012.03979.x.

20. J. D. Rothstein et al., "Beta-lactam antibiotics offer neuroprotection by increasing glutamate transporter expression," *Nature* 433 (2005): 73–77, doi:10.1038/nature03180.

21. B. L. Hart, "Biological basis of the behavior of sick animals," *Neuroscience and Biobehavior Reviews* 12 (1988): 123–137, doi:10.1016/S0149-7634(88)80004-6.

22. S. V. Kasl, A. S. Evans, and J. C. Niederman, "Psychosocial risk factors in the development of infectious mononucleosis," *Psychosomatic Medicine* 41 (1979): 445–466.

23. J. B. Imboden, A. Canter, and E. C. Leighton, "Convalescence from influenza: A study of the psychological and clinical determinants," *Archives*

of Internal Medicine 108 (1961): 393–399, doi:10.1001/archinte.1961.0362009 0065008.

24. R. B. Stricker, "Counterpoint: Long-term antibiotic therapy improves persistent symptoms associated with Lyme disease," *Clinical Infectious Diseases* 45 (2007): 149–157, doi:10.1086/518853.

25. G. P. Wormser and I. Schwartz, "Antibiotic treatment of animals infected with *Borrelia burgdorferi*," *Clinical Microbiological Reviews* 22 (2009): 387–395, doi:10.1128/cmr.00004-09.

26. L. K. Bockenstedt and J. D. Radolf, "Xenodiagnosis for posttreatment Lyme disease syndrome: resolving the conundrum or adding to it?" *Clinical Infectious Diseases* 58 (2014): 946–948, doi:10.1093/cid/cit942.

27. A. Barbour, "Remains of infection," *Journal of Clinical Investigation* 122 (2012): 2344, doi:10.1172/JCI63975.

28. A. Marques et al., "Xenodiagnosis to detect *Borrelia burgdorferi* infection: A first-in-human study," *Clinical Infectious Diseases* 58 (2014): 937–945, doi:10.1093/cid/cit939.

29. L. K. Bockenstedt et al., "Spirochete antigens persist near cartilage after murine Lyme borreliosis therapy," *Journal of Clinical Investigation* 122 (2012): 2652–2660, doi:10.1172/jci58813.

30. N. K. Vudattu et al., "Dysregulation of CD4$^+$ CD25high T cells in the synovial fluid of patients with antibiotic-refractory Lyme arthritis," *Arthritis and Rheumatism* 65 (2013): 1643–1653, doi:10.1002/art.37910.

31. W. W. Spink, "What is chronic brucellosis?" *Annals of Internal Medicine* 35 (1951): 358–374, doi:10.7326/0003-4819-35-2-358.

32. H. J. Harris, "Brucellosis: Advances in diagnosis and treatment," *Journal of the American Medical Association* 131 (1946): 1485–1493, doi:10.1001/jama.1946.02870350017005.

33. W. W. Spink, *The Nature of Brucellosis* (Minneapolis: University of Minnesota, 1956).

34. H. J. Harris and B. L. Stevenson, *Brucellosis (Undulant Fever): Clinical and Subclinical* (New York: Paul B. Hoeber, 1941).

35. Anonymous, "Medicine. Fever from milk," *Time*, May 26, 1941, 45–46.

Chapter 11. Deer Ticks Transmit Other Diseases

1. A. G. Barbour, *Lyme Disease: The Cause, The Cure, The Controversy* (Baltimore, MD: Johns Hopkins University Press, 1996).

2. D. Fish, Boston, MA, 2013, personal communication.

3. J. S. Bakken et al., "Human granulocytic ehrlichiosis in the upper Midwest United States: A new species emerging?" *Journal of the American Medical Association* 272 (1994): 212–218, doi:10.1001/jama.1994.0352003 0054028.

4. E. Hodzic et al., "Acquisition and transmission of the agent of human granulocytic ehrlichiosis by *Ixodes scapularis* ticks," *Journal of Clinical Microbiology* 36 (1998): 3574–3578, www.ncbi.nlm.nih.gov/pmc/articles /PMC105242/.

5. R. J. Thomas, J. S. Dumler, and J. A. Carlyon, "Current management of human granulocytic anaplasmosis, human monocytic ehrlichiosis and *Ehrlichia ewingii* ehrlichiosis," *Expert Reviews in Anti-infective Therapy* 7 (2009): 709–722, doi:10.1586/eri.09.44.

6. Centers for Disease Control and Prevention, *Tickborne Diseases of the United States: A Reference Manual for Health Care Professionals* (Atlanta, GA: U.S. Department of Health and Human Services, 2014), www.cdc.gov /lyme/resources/Tickbornediseases.pdf.

7. T. Ruebush et al., "Human babesiosis on Nantucket Island: Evidence for self-limited and subclinical infections," *New England Journal of Medicine* 297 (1977): 825–827, doi:10.1056/NEJM197710132971511.

8. M. E. P. Acosta et al., "*Babesia microti* infection, eastern Pennsylvania, USA," *Emerging Infectious Diseases* 19 (2013): 1105, doi:10.3201/eid1907.121593.

9. T. Smith and F. L. Kilborne, *Investigations into the Nature, Causation, and Prevention of Texas or Southern Cattle Fever* (Washington, DC: Government Printing Office, 1893).

10. T. N. Mather et al., "*Borrelia burgdorferi* and *Babesia microti*: Efficiency of transmission from reservoirs to vector ticks (*Ixodes dammini*)," *Experimental Parasitology* 70 (1990): 55–61, doi:10.1016/0014-4894(90)90085-Q.

11. B. L. Herwaldt et al., "Transfusion-associated babesiosis in the United States: A description of cases," *Annals of Internal Medicine* 155 (2011): 509–519, doi:10.7326/0003-4819-155-8-201110180-00362.

12. P. J. Krause et al., "Increasing health burden of human babesiosis in endemic sites," *American Journal of Tropical Medicine Hygiene* 68 (2003): 431–436, www.ajtmh.org/content/68/4/431.full.

13. M. A. Diuk-Wasser et al., "Monitoring human babesiosis emergence through vector surveillance, New England, USA," *Emerging Infectious Diseases* 20 (2014): 225–231, doi:10.3201/eid1302/130644.

14. E. Vannier, B. E. Gewurz and P. J. Krause, "Human babesiosis," *Infectious Diseases Clinics of North America* 22 (2008): 469–488, doi:10.1016/j.idc.2008.03.010.

15. C. A. Lobo et al., *"Babesia*: An emerging infectious threat in transfusion medicine," *PLoS Pathogens* 9 (2013): e1003387, doi:10.1371/journal.ppat.1003387.

16. N. P. Tavakoli et al., "Fatal case of deer tick virus encephalitis," *New England Journal of Medicine* 360 (2009): 2099–2107, doi:10.1056/NEJMoa0806326.

17. M. Y. El Khoury et al., "Potential role of deer tick virus in Powassan encephalitis cases in Lyme disease–endemic areas of New York, U.S.A," *Emerging Infectious Diseases* 19 (2013): 1926–1933, doi:10.3201/eid1912.130903.

18. R. A. Nofchissey et al., "Seroprevalence of Powassan virus in New England deer, 1979–2010," *American Journal of Tropical Medicine and Hygiene* 88 (2013): 1159–1162, doi:10.4269/ajtmh.12-0586.

19. J. F. Anderson and P. M. Armstrong, "Prevalence and genetic characterization of Powassan virus strains infecting *Ixodes scapularis* in Connecticut," *American Journal of Tropical Medicine and Hygiene* 87 (2012): 754–759, doi:10.4269/ajtmh.2012.12-0294.

20. G. D. Ebel et al., "Enzootic transmission of deer tick virus in New England and Wisconsin sites," *American Journal of Tropical Medicine and Hygiene* 63 (2000): 36–42, www.ajtmh.org/content/63/1/36.full.pdf+html.

21. R. Lani et al., "Tick-borne viruses: A review from the perspective of therapeutic approaches," *Ticks and Tick-borne Disease*, doi:10.1016/j.ttbdis.2014.04.001.

22. H. R. Chowdri et al., *"Borrelia miyamotoi* infection presenting as human granulocytic anaplasmosis: A case report," *Annals of Internal Medicine* 159 (2013): 21–27, doi:10.7326/0003-4819-159-1-201307020-00005.

23. Anonymous, "Otto Obermeier (1843–1873)," *Nature* 151 (1943): 194.

24. A. G. Barbour, "Relapsing fever," in *Harrison's Principles of Internal Medicine*, 19th ed., ed. D. L. Kasper, S. L. Hauser, D. L. Long, J. L. Jameson, and J. Loscalzo (New York: McGraw Hill, 2015).

25. M. Fukunaga et al., "Genetic and phenotypic analysis of *Borrelia miyamotoi* sp. nov., isolated from the ixodid tick *Ixodes persulcatus*, the vector for Lyme disease in Japan," *International Journal of Systematic Bacteriology* 45 (1995): 804–810, doi:10.1099/00207713-45-4-804.

26. F. Hue, A. Ghalyanchi Langeroudi, and A. G. Barbour, "Chromosome sequence of *Borrelia miyamotoi*, an uncultivable tick-borne agent of human infection," *Genome Announcements* 1 (2013), doi:10.1128/genomeA. 00713-13.

27. A. G. Barbour et al., "Niche partitioning of *Borrelia burgdorferi* and *Borrelia miyamotoi* in the same tick vector and mammalian reservoir species," *American Journal of Tropical Medicine and Hygiene* 81 (2009): 1120–1131, doi:10.4269/ajtmh.2009.09-0208.

28. P. J. Krause et al., "*Borrelia miyamotoi* sensu lato seroreactivity and seroprevalence in the northeastern United States," *Emerging Infectious Diseases* 20 (2014): 1183–1190, doi:10.3201/eid2007.131587.

29. S. J. Swanson et al., "Coinfections acquired from *Ixodes* ticks," *Clinical Microbiology Reviews* 19 (2006): 708–727, doi:10.1128/cmr.00011-06.

30. A. C. Steere et al., "Prospective study of coinfection in patients with erythema migrans," *Clinical Infectious Diseases* 36 (2003): 1078–1081, doi:10.1086/ 368187.

31. P. J. Krause et al., "Disease-specific diagnosis of coinfecting tickborne zoonoses: Babesiosis, human granulocytic ehrlichiosis, and Lyme disease," *Clinical Infectious Diseases* 34 (2002): 1184–1191, doi:10.1086/339813.

32. S. R. Telford III and G. P. Wormser, "*Bartonella* spp. transmission by ticks not established," *Emerging Infectious Diseases* 16 (2010): 379–384, doi:10.3201/eid1603.090443.

33. H. M. Feder et al., "Southern tick-associated rash illness (STARI) in the North: STARI following a tick bite in Long Island, New York," *Clinical Infectious Diseases* 53 (2011): e142–e146, doi:10.1093/cid/cir553

34. E. J. Masters, C. N. Grigery, and R. W. Masters, "STARI, or Masters disease: Lone Star tick-vectored Lyme-like illness," *Infectious Diseases Clinics of North America* 22 (2008): 361–376, viii, doi:10.1016/j.idc.2007.12.010.

35. G. P. Wormser et al., "Prospective clinical evaluation of patients from Missouri and New York with erythema migrans–like skin lesions," *Clinical Infectious Diseases* 41 (2005): 958–965, doi:10.1086/432935.

36. J. E. Childs and C. D. Paddock, "The ascendancy of *Amblyomma americanum* as a vector of pathogens affecting humans in the United States," *Annual Review of Entomology* 48 (2003): 307–337, doi:10.1146/annurev.ento. 48.091801.112728.

37. A. G. Barbour et al., "Identification of an uncultivable *Borrelia* species in the hard tick *Amblyomma americanum*: Possible agent of a Lyme

disease–like illness," *Journal of Infectious Diseases* 173 (1996): 403–409, doi:10.1093/infdis/173.2.403.

38. D. M. Pastula et al., "Notes from the field: Heartland virus disease— United States, 2012–2013," *Morbidity and Mortality Weekly Report* 63 (2014): 270–271, www.cdc.gov/mmwr/preview/mmwrhtml/mm6312a4.htm.

39. D. Brown, "Bush apparently had Lyme disease," *Washington Post,* August 9, 2007, A2.

40. G. E. Griffith et al., *Ecoregions of Texas* (Corvallis, OR: U.S. Environmental Protection Agency, 2004), ftp://ftp.epa.gov/wed/ecoregions/tx/tx_eco _lg.pdf.

41. S. G. Stolberg, "Bush, on a quick trip from his Texas ranch, says Americans are safer than before Sept. 11," *New York Times,* August 11, 2006.

42. S. G. Stolberg, "Letter from Texas: Political reality limits Bush's ranch vacation," *International Herald Tribune,* August 5, 2006.

43. D. T. Dennis et al., "Reported distribution of *Ixodes scapularis* and *Ixodes pacificus* (Acari: Ixodidae) in the United States," *Journal of Medical Entomology* 35 (1998): 629–638.

Chapter 12. Preventing Lyme Disease: Community-Wide Measures

1. S. Hempel, *The Strange Case of the Broad Street Pump: John Snow and the Mystery of Cholera* (Berkeley: University of California Press, 2007).

2. A. G. Barbour and D. Fish, "The biological and social phenomenon of Lyme disease," *Science* 260 (1993): 1610–1616, doi:10.1126/science.8503006.

3. L. L. Williams, Jr., "Malaria eradication in the United States," *American Journal of Public Health and the Nation's Health* 53 (1963): 17–21, ajph.aphapublications.org/doi/pdf/10.2105/AJPH.53.1.17.

4. J. M. Andrews, G. E. Quinby, and A. D. Langmuir, "Malaria eradication in the United States," *American Journal of Public Health and the Nation's Health* 40 (1950): 1405–1411, ajph.aphapublications.org/doi/pdf /10.2105/AJPH.40.11.1405.

5. H. S. Ginsberg, "Integrated pest management and allocation of control efforts for vector-borne diseases," *Journal of Vector Ecology* 26 (2001): 32–38, www.sove.org/Journal/Entries/2001/6/1_Volume_26,_Number_1 _files/ginsberg.pdf.

6. F. D. Guerrero et al., "Acaricide research and development, resistance, and resistance monitoring," in *Biology of Ticks,* ed. D. E. Sonenshine and R. M. Roe (New York: Oxford University Press, 2014), 353–381.

7. E. B. Mitchell et al., "Efficacy of afoxolaner against *Ixodes scapularis* ticks in dogs," *Veterinary Parasitology* 201 (2014): 223–225, doi:10.1016/j.vetpar.2014.02.015.

8. I. Uspensky, "Ticks as the main target of human tick-borne disease control: Russian practical experience and its lessons," *Journal of Vector Ecology* 24 (1999): 40–53.

9. T. L. Schulze et al., "Effects of an application of granular carbaryl on nontarget forest floor arthropods," *Journal of Economic Entomology* 94 (2001): 123–128, doi:10.1603/0022-0493-94.1.123.

10. K. C. Stafford III, *Tick Management Handbook*, rev. ed. (New Haven: Connecticut Agricultural Experiment Station, 2007), stacks.cdc.gov/view/cdc/11444.

11. H. S. Ginsberg and K. C. Stafford III, "Management of ticks and tick-borne diseases," in *Tick-Borne Diseases of Humans*, ed. J. L. Goodman, D. T. Dennis, and D. E. Sonenshine (Washington, DC: ASM Press, 2005), 65–86.

12. H. S. Ginsberg, "Tick control: Trapping, biocontrol, host management, and other alternative strategies," in *Biology of Ticks*, ed. D. E. Sonenshine and R. M. Roe (New York: Oxford University Press, 2014), 409–444.

13. K. C. Stafford III, A. J. Denicola, and H. J. Kilpatrick, "Reduced abundance of *Ixodes scapularis* (Acari: Ixodidae) and the tick parasitoid *Ixodiphagus hookeri* (Hymenoptera: Encyrtidae) with reduction of white-tailed deer," *Journal of Medical Entomology* 40 (2003): 642–652, doi:10.1603/0022-2585-40.5.642.

14. M. Samish and J. Rehacek, "Pathogens and predators of ticks and their potential in biological control," *Annual Review of Entomology* 44 (1999): 159–182, doi:10.1146/annurev.ento.44.1.159.

15. E. Zhioua et al., "Pathogenicity of *Bacillus thuringiensis* variety kurstaki to *Ixodes scapularis* (Acari: Ixodidae)," *Journal of Medical Entomology* 36 (1999): 900–902.

16. R. S. Ranju et al., "Field trials to attract questing stages of brown dog tick, *Rhipicephalus sanguineus,* using tick pheromone-acaricide complex," *Journal of Parasitic Diseases* 37 (2013): 84–87, doi:10.1007/s12639-012-0136-x.

17. P. J. Kelly et al., "Efficacy of slow-release tags impregnated with aggregation-attachment pheromone and deltamethrin for control of *Amblyomma variegatum* on St. Kitts, West Indies," *Parasites and Vectors* 7 (2014): 182, doi:10.1186/1756-3305-7-182.

18. R. P. Smith et al., "Norway rats as reservoir hosts for Lyme disease spirochetes on Monhegan Island, Maine," *Journal of Infectious Diseases* 168 (1993): 687–691, doi:10.1093/infdis/168.3.687.

19. D. Markowski et al., "Spatial distribution of larval *Ixodes scapularis* (Acari: Ixodidae) on *Peromyscus leucopus* and *Microtus pennsylvanicus* at two island sites," *Journal of Parasitology* 83 (1997): 207–211, www.jstor.org/stable /3284440.

20. T. N. Mather, J. M. Ribeiro, and A. Spielman, "Lyme disease and babesiosis: Acaricide focused on potentially infected ticks," *American Journal of Tropical Medicine and Hygiene* 36 (1987): 609–614.

21. D. Fish and J. E. Childs, "Community-based prevention of Lyme disease and other tick-borne diseases through topical application of acaricide to white-tailed deer: Background and rationale," *Vector Borne and Zoonotic Diseases* 9 (2009): 357–364, doi:10.1089/vbz.2009.0022.

22. M. C. Dolan et al., "Control of immature *Ixodes scapularis* (Acari: Ixodidae) on rodent reservoirs of *Borrelia burgdorferi* in a residential community of southeastern Connecticut," *Journal of Medical Entomology* 41 (2004): 1043–1054, doi:10.1603/0022-2585-41.6.1043.

23. R. S. Lane et al., "A better tick-control trap: Modified bait tube controls disease-carrying ticks and fleas," *California Agriculture* 52 (1998): 43–47, doi:10.3733/ca.v052n02p43.

24. J. I. Tsao et al., "An ecological approach to preventing human infection: Vaccinating wild mouse reservoirs intervenes in the Lyme disease cycle," *Proceedings of the National Academy of Science U.S.A.* 101 (2004): 18159–18164, doi:10.1073/pnas.0405763102.

25. R. T. Sterner et al., "Tactics and economics of wildlife oral rabies vaccination, Canada and the United States," *Emerging Infectious Diseases* 15 (2009): 1176–1184, doi:10.3201/eid1508.081061.

26. D. Bhattacharya et al., "Development of a baited oral vaccine for use in reservoir-targeted strategies against Lyme disease," *Vaccine* 29 (2011): 7818–7825, doi:10.1016/j.vaccine.2011.07.100.

27. M. C. Dolan et al., "Elimination of *Borrelia burgdorferi* and *Anaplasma phagocytophilum* in rodent reservoirs and *Ixodes scapularis* ticks using a doxycycline hyclate-laden bait," *American Journal of Tropical Medicine and Hygiene* 85 (2011): 1114–1120, doi:10.4269/ajtmh.2011. 11-0292.

28. S. D. Côté et al., "Ecological impacts of deer overabundance," *Annual Review of Ecology, Evolution, and Systematics* 35 (2004): 113–147, www.jstor .org/stable/30034112.

29. M. L. Wilson et al., "Host-dependent differences in feeding and reproduction of *Ixodes dammini* (Acari: Ixodidae)," *Journal of Medical Entomology* 27 (1990): 945–954.

30. J. E. Childs and C. D. Paddock, "The ascendancy of *Amblyomma americanum* as a vector of pathogens affecting humans in the United States," *Annual Review of Entomology* 48 (2003): 307–337, doi:10.1146/annurev.ento .48.091801.112728.

31. M. R. Conover, "Monetary and intangible valuation of deer in the United States," *Wildlife Society Bulletin* 25 (1997): 298–305.

32. P. W. Rand et al., "Abundance of *Ixodes scapularis* (Acari: Ixodidae) after the complete removal of deer from an isolated offshore island, endemic for Lyme disease," *Journal of Medical Entomology* 41 (2004): 779–784, doi:10.1603/0022-2585-41.4.779.

33. J. F. Kirkpatrick, R. O. Lyda, and K. M. Frank, "Contraceptive vaccines for wildlife: a review," *American Journal of Reproductive Immunology* 66 (2011): 40–50, doi:10.1111/j.1600-0897.2011.01003.x.

34. R. Salit, "Block Island culling deer herd with hunting-by-lottery, bait-and-shoot effort," *Providence Journal* (RI), January 24, 2014.

35. J. M. Pound et al., "The '4-poster' passive topical treatment device to apply acaricide for controlling ticks (Acari: Ixodidae) feeding on white-tailed deer," *Journal of Medical Entomology* 37 (2000): 588–594, doi:10.1603/0022- 2585-37.4.588.

36. B. Brei et al., "Evaluation of the United States Department of Agriculture Northeast Area-Wide Tick Control Project by meta-analysis," *Vector Borne and Zoonotic Diseases* 9 (2009): 423–430, doi:10.1089/vbz. 2008.0150.

37. P. Boody, "Pesticide used in 4-poster tick-killing, deer-feeding stations now legal on Long Island," *Shelter Island Reporter* (NY), January 28, 2012.

38. J. de la Fuente et al., "A ten-year review of commercial vaccine performance for control of tick infestations on cattle," *Animal Health Research Reviews* 8 (2007): 23–28, doi:10.1017/s1466252307001193.

39. K. LoGiudice et al., "The ecology of infectious disease: Effects of host diversity and community composition on Lyme disease risk," *Proceedings of*

the National Academy of Science U.S.A. 100 (2003): 567–571, doi:10.1073/pnas. 0233733100.

40. N. H. Ogden and J. I. Tsao, "Biodiversity and Lyme disease: Dilution or amplification?" *Epidemics* 1 (2009): 196–206, doi:10.1016/j.epidem. 2009.06.002.

41. S. States et al., "Lyme disease risk not amplified in a species-poor vertebrate community: Similar *Borrelia burgdorferi* tick infection prevalence and OspC genotype frequencies," *Infection, Genetics and Evolution* 27 (2014): 566–575, doi:10.1016/j.meegid.2014.04.014.

42. Environmental Protection Agency, "Integrated pest management (IPM) principles," updated August 5, 2014, www.epa.gov/opp00001 /factsheets/ipm.htm.

43. T. L. Schulze et al., "Integrated use of 4-poster passive topical treatment devices for deer, targeted acaricide applications, and Maxforce TMS bait boxes to rapidly suppress populations of *Ixodes scapularis* (Acari: Ixodidae) in a residential landscape," *Journal of Medical Entomology* 44 (2007): 830–839, doi:10.1603/0022-2585(2007)44[830:IUOPPT]2.0.CO;2.

Chapter 13. Preventing Lyme Disease: Personal and Household Protection

1. L. H. Daltroy et al., "A controlled trial of a novel primary prevention program for Lyme disease and other tick-borne illnesses," *Health Education and Behavior* 34 (2007): 531–542, doi:10.1177/1090198106294646.

2. D. Richter et al., "How ticks get under your skin: Insertion mechanics of the feeding apparatus of *Ixodes ricinus* ticks," *Proceedings of the Royal Society B: Biological Sciences* 280 (2013), doi:10.1098/rspb.2013.1758.

3. S. Warshafsky et al., "Efficacy of antibiotic prophylaxis for the prevention of Lyme disease: An updated systematic review and meta-analysis," *Journal of Antimicrobial Chemotherapy* 65 (2010): 1137–1144, doi:10.1093/jac/dkq097.

4. G. P. Wormser et al., "The clinical assessment, treatment, and prevention of Lyme disease, human granulocytic anaplasmosis, and babesio-sis: Clinical practice guidelines by the Infectious Diseases Society of America," *Clinical Infectious Diseases* 43 (2006): 1089–1134, doi:10.1086/508667.

5. J. Piesman et al., "Efficacy of an experimental azithromycin cream for prophylaxis of tick-transmitted Lyme disease spirochete infection in a murine model," *Antimicrobial Agents and Chemotherapy* 58 (2014): 348–351, doi:10.1128/aac.01932-13.

6. M. S. Fradin and J. F. Day, "Comparative efficacy of insect repellents against mosquito bites," *New England Journal of Medicine* 347 (2002): 13–18, doi:10.1056/NEJMoa011699.

7. F. Pages et al., "Tick repellents for human use: Prevention of tick bites and tick-borne diseases," *Vector Borne and Zoonotic Diseases* 14 (2014): 85–93, doi:10.1089/vbz.2013.1410.

8. A. Gardulf, I. Wohlfart, and R. Gustafson, "A prospective cross-over field trial shows protection of lemon eucalyptus extract against tick bites," *Journal of Medical Entomology* 41 (2004): 1064–1067, doi:10.1603/0022-2585-41.6.1064.

9. D. Strickman, S. P. Frances, and M. Debboun, *Prevention of Bug Bites, Stings, and Disease* (New York: Oxford University Press, 2009).

10. V. Chen-Hussey, R. Behrens, and J. G. Logan, "Assessment of methods used to determine the safety of the topical insect repellent N,N-diethyl-m-toluamide (DEET)," *Parasites and Vectors* 7 (2014): 173, doi:10.1186/1756-3305-7-173.

11. Environmental Protection Agency, "3-[N-Butyl-N-acetyl]-aminopropi onic acid, ethyl ester (IR3535) (113509) Fact Sheet," January 1, 2000, www.epa .gov/opp00001/chem_search/reg_actions/registration/fs_PC-113509_01-Jan -00.pdf.

12. E. Lupi, C. Hatz, and P. Schlagenhauf, "The efficacy of repellents against *Aedes, Anopheles, Culex* and *Ixodes* spp.—a literature review," *Travel Medicine and Infectious Diseases* 11 (2013): 374–411, doi:10.1016/j.tmaid.2013. 10.005.

13. B. W. Bissinger et al., "Comparative efficacy of BioUD to other commercially available arthropod repellents against the ticks *Amblyomma americanum* and *Dermacentor variabilis* on cotton cloth," *American Journal of Tropical Medicine and Hygiene* 81 (2009): 685–690, doi:10.4269/ajtmh.2009. 09-0114.

14. W. L. Shoop et al., "Discovery and mode of action of afoxolaner, a new isoxazoline parasiticide for dogs," *Veterinary Parasitology* 201 (2014): 179–189, doi:10.1016/j.vetpar.2014.02.020.

15. E. B. Mitchell et al., "Efficacy of afoxolaner against *Ixodes scapularis* ticks in dogs," *Veterinary Parasitology* 201 (2014): 223–225, doi:10.1016/j.vetpar. 2014.02.015.

16. D. E. Snyder, L. R. Cruthers, and R. L. Slone, "Preliminary study on the acaricidal efficacy of spinosad administered orally to dogs infested with

the brown dog tick, *Rhipicephalus sanguineus* (Latreille, 1806) (Acari: Ixodidae)," *Veterinary Parasitology* 166 (2009): 131–135, doi:10.1016/j.vetpar. 2009.07.046.

17. D. Fish and J. E. Childs, "Community-based prevention of Lyme disease and other tick-borne diseases through topical application of acaricide to white-tailed deer: Background and rationale," *Vector Borne and Zoonotic Diseases* 9 (2009): 357–364, doi:10.1089/vbz.2009.0022.

18. H. S. Ginsberg and K. C. Stafford III, "Management of ticks and tick-borne diseases," in *Tick-Borne Diseases of Humans*, ed. J. L. Goodman, D. T. Dennis, and D. E. Sonenshine (Washington, DC: ASM Press, 2005), 65–86.

19. K. C. Stafford III, *Tick Management Handbook*, rev. ed. (New Haven: Connecticut Agricultural Experiment Station, 2007), www.ct.gov/caes/lib /caes/documents/publications/bulletins/b1010.pdf.

20. A. G. Barbour, S. L. Tessier, and W. J. Todd, "Lyme disease spiro-chetes and ixodid tick spirochetes share a common surface antigenic determi-nant defined by a monoclonal antibody," *Infection and Immunity* 41 (1983): 795–804, www.ncbi.nlm.nih.gov/pmc/articles/PMC264710/.

21. E. Fikrig et al., "Protection of mice against the Lyme disease agent by immunizing with recombinant OspA," *Science* 250 (1990): 553–556, doi:10.1126/science.2237407.

22. L. Nigrovic and K. Thompson, "The Lyme vaccine: A cautionary tale," *Epidemiology and Infection* 135 (2007): 1–8, doi:10.1017/S0950268806007096.

23. C. Willyard, "Resurrecting the 'yuppie vaccine,'" *Nature Medicine* 20 (2014): 698–701, doi:10.1038/nm0714-698.

24. C. Nickisch, "Why your dog can get vaccinated against Lyme disease and you can't," *WBUR,* Boston's NPR news station, June 27, 2012, www.wbur .org/2012/06/27/lyme-vaccine.

25. S. A. Plotkin, "Bring back the Lyme vaccine," *New York Times,* September 18, 2013.

26. G. A. Poland, "Vaccines against Lyme disease: What happened and what lessons can we learn?" *Clinical Infectious Diseases* 52, supplement 3 (2011): s253–s258, doi:10.1093/cid/ciq116.

27. L. H. Sigal et al., "A vaccine consisting of recombinant *Borrelia burgdor-feri* outer-surface protein A to prevent Lyme disease. Recombinant Outer-Surface Protein: A Lyme Disease Vaccine Study Consortium," *New England Journal of Medicine* 339 (1998): 216–222, doi:10.1056/NEJM199807233390402.

28. J. A. Conlon et al., "Efficacy of a nonadjuvanted, outer surface protein A, recombinant vaccine in dogs after challenge by ticks naturally infected with *Borrelia burgdorferi*," *Veterinary Therapy* 1 (2000): 96–107.

29. S. A. Levy, B. A. Lissman, and C. M. Ficke, "Performance of a *Borrelia burgdorferi* bacterin in borreliosis-endemic areas," *Journal of the American Veterinary Medical Association* 202 (1993): 1834–1838.

30. A. C. Steere et al., "Vaccination against Lyme disease with recombinant *Borrelia burgdorferi* outer-surface lipoprotein A with adjuvant: Lyme Disease Vaccine Study Group," *New England Journal of Medicine* 339 (1998): 209–215, doi:10.1056/nejm199807233390401.

31. M. P. Littman et al., "ACVIM small animal consensus statement on Lyme disease in dogs: Diagnosis, treatment, and prevention," *Journal of Veterinary Internal Medicine* 20 (2006): 422–434, www.cvm.ncsu.edu/vhc /documents/LymeconsstmtACVIM.pdf.

32. M. Stone, "Is it a good idea for dogs to get the canine Lyme vaccine, even if they already had Lyme disease?" June 9, 2012, http://now.tufts.edu /articles/dogs-canine-lyme-vaccine.

33. R. N. Basu, Z. Jezek, and N. A. Ward, *The Eradication of Smallpox from India* (New Delhi, India: World Health Organization, 1979).

Glossary

acaricide: A chemical compound or biological material that is injurious to ticks and mites (chapter 12). Acaricides at present are mostly drawn from the broader category of insecticides.

adaptive immunity: The immune response of humans and other vertebrates that develops after exposure to a pathogen and that is more specifically adapted to the pathogen than *innate immunity* generally is (chapter 1). Antibodies are a manifestation of adaptive immunity and are the main defense against *Borrelia* bacteria.

biological control: The management of arthropod and other pests, such as disease *vectors*, with living organisms (chapter 12). These controls range from fish that are predators of mosquito larvae to bacteria and fungi that are natural pathogens of the pest.

blood smear: A drop of blood thinly smeared on a glass slide, fixed in place with alcohol, and then treated with a stain or antibodies to reveal the suspected pathogen. Used for the detection of *Babesia microti, Anaplasma phagocytophilum,* and *Borrelia miyamotoi* in the blood (chapter 11), but of no use for detecting Lyme disease *Borrelia.*

Borrelia: A genus in the *spirochete* division of bacteria. All known *Borrelia* species have one or more vertebrates as a *reservoir* and an arthropod, usually a tick, as a *vector.* Examples of *Borrelia* species are *Borrelia burgdorferi,* a Lyme disease species, and *B. miyamotoi,* a relapsing fever species. (After its first use in a text, the genus name is usually abbreviated to a single letter; here, *B.*)

case definition: A set of epidemiological, clinical, and/or laboratory criteria for counting a particular illness as a case of a particular disease or syndrome.

Case definitions are used for epidemiology and record-keeping purposes and are not necessarily the same as diagnoses, which may be made on less certain grounds and are meant to guide subsequent therapy and management.

CDC: Abbreviation for the Centers for Disease Control and Prevention, an agency of the United States Department of Health and Human Services. The home office and its laboratories are located in Atlanta, Georgia. The branch that deals with Lyme disease is located in Fort Collins, Colorado. The CDC is mainly concerned with the epidemiology and prevention of diseases and has a more limited research role than its sister institution in the Public Health Service, the *NIH*.

cephalosporins: A class of antibiotics related to *penicillins* that work by interfering with the formation of the bacterial cell's wall (chapter 8). There are many types of cephalosporins; some are effective against *Borrelia,* and some are not. The ones most commonly used for treatment of Lyme disease are cefuroxime for oral use and ceftriaxone for intravenous use. Most cephalosporins can be taken by young children.

chronic Lyme disease: The term originally synonymous with *late Lyme disease,* that is, an untreated symptomatic infection with *B. burgdorferi* of several weeks or longer duration. By current usage, however, the term is applied by some practitioners to cases in which symptoms and disability persist after the usual course of antibiotic treatment. The implications are that spirochetes are still alive and that much longer or different treatments are called for. In this context, *chronic Lyme disease* is distinguished from *late Lyme disease* and *post-treatment Lyme disease syndrome* (chapter 10).

cross-reactivity: In reference to antibodies, this term or its adjective, cross-reactive, means that an antibody originally elicited by infection with one type of microbe may also bind to other microbes or their parts because of similarities in their substances. Cross-reactions of this type can lead to false-positives in laboratory tests like the *EIA*.

early Lyme disease: The first month or two of *B. burgdorferi* infection (chapter 1). Common manifestations are the *erythema migrans* skin rash and flu-like symptoms of fever, fatigue, muscle aches, and headache. Some other manifestations that generally appear after erythema migrans starts are multiple skin rashes, facial palsy (weakness of one or both sides of the facial muscles), and generalized joint pains. Early disease may be localized, as occurs when the infection is limited to the area around the tick bite site, or disseminated, as occurs when the pathogens invade the blood. There is no hard-and-fast time boundary between early and *late Lyme disease.*

EIA: An abbreviation for enzyme immunoassay, a laboratory test for detecting and measuring the amount of antibodies to a particular microbe in a vertebrate's blood or cerebrospinal fluid through use of an enzyme to quantify the amount of bound antibody (chapter 6). The alternative name for this test is ELISA (enzyme-linked immunosorbent assay).

erythema migrans: A localized, sometimes patterned, reddish skin rash that commonly but not invariably occurs at the site where a tick inoculates Lyme disease spirochetes in the skin of a human host (chapter 1). If the spirochetes disseminate in the blood, there may be multiple erythema migrans lesions at distant sites. Once thought to be largely specific for Lyme disease, a similar rash may occur with southern tick-associated rash illness, or STARI (chapter 11).

host: In the context of this book, a vertebrate, such as the white-footed mouse, that *B. burgdorferi* or other microbe infects and that occupies one stage of the life cycle for the pathogen. It also refers to the vertebrates, such as deer, that the ticks feed on. Strictly speaking, a tick is also a host for *B. burgdorferi*, but the term *vector* is used instead for the tick.

IFA: Abbreviation for indirect immunofluorescence assay, a laboratory test used for detecting and measuring the amount of antibodies to a particular microbe in the serum of a vertebrate. The amount of antibody bound is quantified through use of a second antibody that has a fluorescent tag. The fluorescence is visualized by microscopy under ultraviolet light. IFA is uncommonly used for Lyme disease diagnosis today but is still frequently used for laboratory tests for babesiosis and human granulocytic anaplasmosis.

incidence: The frequency of new cases of a disease or condition during a given period, usually a year. The disease may last for just a few days or for the entire period, but it would only count once. Incidence is different from *prevalence.*

infestation: In the context of this book, the state of having ticks or other arthropods occupying one's skin, hair, or clothes. The term usually applies to external parasites, like ticks, mites, lice, or fleas, and so is generally distinguished from infection, which applies to internal parasites and pathogens. There may be infestation of an animal, and this may be of some consequence, such as blood loss or itching, but there is not necessarily transmission of an infection.

innate immunity: The combination of defenses that an animal, including invertebrates, mounts in response to infection (chapter 1). In general, the responses do not finely discriminate between pathogens and in this way are distinguishable from *adaptive immunity*. Innate immunity is the first response to infection and often leads to inflammation.

Ixodes: The genus name for the group of tick species that include the *vectors* of *B. burgdorferi* and other pathogens to humans and other animals (chapter 3). The name derives from the Greek word for "sticky." There are scores of *Ixodes* species, and one or another of these species can be found on all continents. The species that transmit Lyme disease are *I. scapularis* and *I. pacificus* in North America and *I. ricinus* and *I. persulcatus* in Eurasia.

late Lyme disease: A symptomatic active infection with *B. burgdorferi* (or related species) lasting roughly for two months or more and revealing itself in one or more different ways, such as arthritis of one or a few large joints (Lyme arthritis), disorders of the nerves or brain, or, in Europe alone, a persistent inflammatory condition of the skin (chapter 2). There is the presumption, if not proof, that live spirochetes are present and, furthermore, that these spirochetes can usually be eliminated with antibiotic treatment of a no more than few weeks' duration. This term contrasts with current usage of the term *chronic Lyme disease*, for which there is not the general presumption that antibiotic treatment of this duration will be successful. Admittedly, this is an imprecise and blurred distinction, which is the source of confusion for patients and doctors alike (introduction, chapter 2, and chapter 10).

Lyme borreliosis: An alternative term for *Lyme disease*, and one more in common use in Europe than in North America (introduction).

Lyme disease: A symptomatic, *zoonotic* infection caused by *B. burgdorferi* in North America and by either *B. afzelii* or *B. garinii* in Eurasia. This infection is acquired from an *Ixodes* tick *vector*.

macrolides: A class of antibiotics that share a similar chemical structure and mechanism of action: interfering with protein synthesis by bacteria. The two types that are sometimes used for treatment of Lyme disease are erythromycin and azithromycin.

NIH: Abbreviation for the National Institutes of Health, an agency of the United States Department of Health and Human Services. The NIH is more focused on basic and clinical research than on epidemiology and public health service, which are more the domain of the *CDC*. The NIH, whose main offices and laboratories are located in Bethesda, Maryland, supports most of the research on Lyme disease in the United States. One of the NIH's subdivisions, the Rocky Mountain Laboratories in Hamilton, Montana, is where the cause of Lyme disease was discovered (introduction).

PCR: Abbreviation for polymerase chain reaction, the laboratory method to amplify in repetitive steps a small fragment of DNA of a microbe millions and millions of times with a special enzyme, to the point that it can be easily measured (chapter 6). PCR is a method of direct detection of the *B. burgdorferi* and other microbes, but it does not distinguish between live and dead cells.

penicillins: A class of antibiotics that are related to the cephalosporins and that work by interfering with the formation of the bacterial cell's wall (chapter 8). There are many types of penicillins. The one most commonly used for treatment of Lyme disease is amoxicillin, but the original formulation of penicillin is also effective, albeit less convenient in its dosing (chapter 9). Most penicillins can be taken by young children.

post-treatment Lyme disease syndrome: The coherent collection of otherwise unexplained persistent or recurrent symptoms after completion of antibiotic therapy for Lyme disease and for which further antibiotic therapy does not seem to provide benefit (chapter 10). Inherent in the term is the presumption that living cells of *B. burgdorferi* are no longer present and that the syndrome has another explanation than continued infection per se. The abbreviation is PTLDS. An alternative term for the condition is "post–Lyme disease syndrome." A particular patient might be said to have PTLDS by one physician and *chronic Lyme disease* by another physician, depending on the physician's point of view (introduction).

prevalence: The presence of a disease or condition in a population at a particular snapshot in time. The disease may have begun in the past—that is, during another period of measurement, such as 5 years previously—but it is still present at the current time of assessment. So it is counted as present in each of the 5 years. The term is also used for the proportion of ticks or small mammals that are infected at the time of their collection. Prevalence is different from *incidence*. As an example, the *incidence* of new infections with a virus might be 40 per 100,000 people in a given year, but the *prevalence* of infections with the virus, which, let's suppose, is a chronic, incurable condition, is 400 per 100,000. The number 400 includes all the individuals currently living with infections with the virus, no matter when it was first acquired.

prophylaxis: The prevention of disease with a treatment, either before exposure, such as with a medicine against malaria before entering a risk area, or just after exposure, such as by taking an antibiotic after a tick bite (chapter 13).

reservoir: The animal that is the usual or common host of a *zoonotic* pathogen (chapter 3). Reservoir hosts may be seriously sickened by the infection but more commonly are not affected by the microbe or suffer only mild illnesses.

spirochete: A type of bacterium with a unique cell structure and mode of locomotion (chapter 3). On an evolutionary scale, spirochetes are as different from some other bacteria, like *E. coli* or the staph bacterium, as humans are from pine trees. There are hundreds of species of spirochetes, only a handful of which cause diseases of humans.

tetracyclines: A class of antibiotics that share a similar chemical structure (chapter 8). Like macrolides, tetracyclines interfere with protein synthesis by the bacterium but by a different mechanism. Although the only tetracycline that is used at present for treatment of Lyme disease is doxycycline, earlier formulations could be substituted if there was a shortage of doxycycline. Tetracyclines are contraindicated for children younger than 9 years of age and for pregnant or nursing women, because of possible adverse effect on a fetus's or a child's teeth.

vector: In the context of the book, a vector is the tick that transmits the *Borrelia* species or other microbe from one vertebrate *reservoir* to another or, occasionally, to an inadvertent *host*, like a human or dog (chapters 3 and 4). For the tick to be a competent vector, the microbe has to be able to pass from one stage of the life cycle to another, for example, from larval to nymphal stage, after feeding on one vertebrate and then pass out of the tick into another vertebrate host.

Western blot: A laboratory assay that usually follows an *EIA* in a two-step testing protocol for Lyme disease diagnosis (chapter 6). The Western blot generally provides better specificity than an EIA, because it is easier to distinguish between *cross-reactive* substances and those that are restricted to the pathogen of interest. Western blot tests based on broken-up cells of *B. burgdorferi* are gradually being supplanted by assays that are based on mixtures of purified proteins.

zoonosis: An infectious disease transmitted naturally to humans from another animal or animals, either directly, such as for brucellosis (chapter 10), or indirectly through a *vector*, such as for Lyme disease (chapter 3) or babesiosis (chapter 11).

Index

Acaricides: area-wide application of, 231–232; definition of, 226–227; resistance to, 230–231; types, 227–230. *See also under name of specific acaricide or insecticide*

Acrodermatitis chronica atrophicans, 46, 96, 120

Adaptive immunity. *See* Immunity: adaptive

Afoxolaner insecticide, 229, 260

Afzelius, Arvid, 1–2, 5, 18

Allergy to antibiotics. *See* Antibiotics: allergic reactions to

Amblyomma americanum (Lone star tick): association of, with STARI, 216–217; *Borrelia lonestari* and, 217; geographic distribution of, 217–218, 249; heartland virus disease and, 218; human monocytic ehrlichiosis and, 218

American Lyme Disease Foundation, 271

Aminoglycoside antibiotics, 148

Amitraz insecticide, 228, 259–260

Amoxicillin, 140, 145, 178, 253

Amyotrophic lateral sclerosis and Lyme disease, 29, 40, 191

Anaplasma phagocytophilum. *See* Human granulocytic anaplasmosis

Antibiotics: adverse (side) effects of, 161–164; allergic reactions to, 165–167; clinical trials of, 151–154, 156–157; combinations of, 171–172; definition of, 141–142; doses and intervals for, 169, 179; duration of treatment with, 154–156, 170, 179; entry into brain by, 144; entry into cells by, 144; interactions with other drugs, 165; intravenous delivery of (*see* Intravenous antibiotic therapy); mechanisms of action of, 142–143; oral delivery of, 143; penetration to brain and CSF by, 144; prophylaxis after tick bites with, 252–254; resistance to, 146–147, 157–158; safety of, 159–161; testing of, in laboratory animals, 149–150; unintended effects of, 189–192; use of, during pregnancy and breast-feeding, 140, 168, 180; use of, for children, 168, 180; wildlife-targeting for, 239–240. *See also under name of specific antibiotic*

Antibodies: cross-reactions of, 105, 196, 202, 214; false-positive reactions with, 39, 109–111, 119; IgG, 115, 131; IgM, 115–116, 119, 131, 138; time course for, 115; types of, 113. *See also* Immunity: adaptive

Antigens: cross-reactive proteins as, 108, 113, 214; definition of, 133n

Arthritis. *See* Joints, infection of

Asymptomatic infection, 24, 111, 172–174

Atovaquone for babesiosis, 208

Autism and Lyme disease, 40
Autoimmunity. *See* Joints, infection of: autoimmunity in
Azithromycin: for babesiosis, 208; for Lyme disease, 147, 154, 175, 179, 254; skin cream (topical), 254. *See also* Macrolide antibiotics

Babesiosis: agents of, 204–205; asymptomatic infection with, 208–209; *Babesia microti* and, 204; diagnosis of, 206–208; in Europe, 204; in far western United States, 204; incidence of, 205; manifestations of, 203–206; prevalence in ticks, 205; reservoirs of, 205; spleen absence and, 206; transfusion transmission of, 206; treatment of, 208–209
Bait tubes (insecticide) for tick control, 237
Bartonella infection risk, 215–216
Bell's palsy (facial paralysis), 32, 116, 155, 173, 179
Benach, Jorge, 124–125
Biological control of ticks: *Bacillus thuringiensis* (Bt) for, 234; definition of, 232; fungi for, 234; wasps for, 233; worms for, 233
Biopsy of skin and other tissues, 7, 34, 120–122
Birds, 55–58, 66–67, 69, 75, 200, 205
Blacklegged tick. See *Ixodes scapularis*
Block Island, Rhode Island, 81, 205, 240, 243
Blood smear, use of, for diagnosis of: babesiosis, 203, 207–208; *Borrelia miyamotoi* infection, 214; human granulocytic anaplasmosis, 201
Borderline and indeterminate test reactions, 39, 109, 131
Borrelia, other species of, that cause Lyme disease: *Borrelia afzelii*, 46, 132; *Borrelia garinii*, 46, 132; *Borrelia lusitaniae*, 46n;

Borrelia spielmanii, 46n; differences in disease manifestation for, 46; geographic distributions of, 45–46
Borrelia burgdorferi: antibiotic susceptibility of, 147–148; climate and landscape for, 69–71; cultivation of, 118–121; discovery of, 2, 4–5; geographic distribution of, 69–75; life cycle of, 65–69; nonprotein antigens of, 109; prevalence in ticks, 198; resistance, lack of, in, 157–158; spread of, 71–75; strains of (*see* Strains of *B. burgdorferi*); virulence of, 48
Borrelia miyamotoi, infection with: agent of, 213; diagnosis of, 213; discovery of, 213; geographic distribution of, 213; incidence of, 214; manifestations of, 214; prevalence in ticks, 213; reservoirs of, 213; treatment of, 213
Brain, infection of. *See* Nervous system, infection of
Breast feeding, Lyme disease during, 168, 180
Brucellosis, 195–197
Burgdorfer, Willy, ix, 124–125

C6 peptide test, 108–109, 174
California, tickborne diseases in, 4, 13, 50, 54, 57, 72, 77–78, 94, 132, 231, 237
Canada, Lyme disease in, 70, 73, 92, 94, 263, 272
Candida yeasts and candidiasis, 140, 164
Carbamate insecticides, 227–228, 260
Carbaryl, 227–228, 257, 259
Cardiac pacemaker. *See* Heart, infection of: treatment of
Carditis, Lyme. *See* Heart, infection of
Case histories: Case A, 1, 4, 11; Case B, 2, 4; Case C, 5; Case D, 12–14; Case E, 12–13; Case F, 12–13, 22; Case G, 27–28, 30–31, 33; Case H, 27–28, 33–34; Case I, 27–28, 37; Case J, 140, 150, 168; Case K,

140–141, 161, 164, 189; Case L, 182–183, 187; Case M, 199; Case N, 203; Case O, 203–204, 208; Case P, 209–210; Case Q, 212; Case R, 216

Case reports and reporting: case definition for, 79–80; collection of, 80; differences of, from clinical diagnosis, 137–138; reportable diseases for, 63; underestimation of, 80–81

Cats: *Borrelia burgdorferi* infection of, 41, 59, 175, 268; tick control for, 230, 260; toxicity of amitraz and pyrethroids for, 230, 260; vaccine, lack of, for, 268

CDC. *See* Centers for Disease Control and Prevention

Cefotaxime, 155. *See also* Cephalosporin antibiotics

Ceftin. *See* Cefuroxime

Ceftriaxone: administration of, 145, 162, 164; adverse effects of, 162–164; effectiveness of, 155, 181, 188–189, 214; unintended effects of, 191–192. *See also* Cephalosporin antibiotics

Cefuroxime, 145, 147, 153–157, 178, 180. *See also* Cephalosporin antibiotics

Cellulitis, 15, 17, 78, 94, 180

Centers for Disease Control and Prevention (CDC), 37, 61, 79, 81, 114, 127, 183, 203, 205, 209, 247–248, 257, 263, 271

Central nervous system (CNS). *See* Nervous system, infection of

Cephalexin (Keflex), 148

Cephalosporin antibiotics: effectiveness of, against *Borrelia*, 147, 150; ineffectiveness of, for HGA, 202; mechanism of action of, 147; non-antimicrobial effects of, 190–191; similarity to penicillins, 147; treatment with, 155–157, 178–179; types of, 147–148. *See also under name of specific antibiotic*

Cerebrospinal fluid (CSF), analysis of, 117–118, 120, 179, 211

Chelation therapy, 175

Children: arthritis in, 3, 34; babesiosis in, 206; erythema migrans sites for, 18; human granulocytic anaplasmosis in, 200, 202–203; representation of, in clinical trials, 168; risk of Lyme disease in, 76–77; treatment of, with tetracyclines, 168, 180; use of tick repellents for, 258–259

Chipmunks, 57, 59, 67, 200, 205, 237, 239, 261

Chloramphenicol, 147–148

Chronic fatigue syndrome, 183, 185–187

"Chronic Lyme disease," 85, 140, 167, 175, 178, 183, 192, 196

Ciprofloxacin, 148

Citrodiol. *See* Repellents: PMD

Clarithromycin, 154. *See also* Macrolide antibiotics

Clavulanate, use of, with amoxicillin, 180

Clindamycin for babesiosis, 208

Clostridium difficile and diarrhea, 163–164

Clothing. *See* Personal and household protection

Coinfections, 215–216

Complications of *B. burgdorferi* infection: affecting heart (*see* Heart, infection of); affecting joints (*see* Joints, infection of); affecting nervous system (*see* Nervous system, infection of); affecting other organs (*see* Lyme disease: other symptoms and complications of)

Connecticut Agricultural Experiment Station, 272

Controlled clinical trials, 160–161

Corticosteroids and Lyme disease treatment, 176

Cutoff point, determination of, in laboratory tests, 109–111, 130–131, 134

Cytokines, 16–17, 192–193

Damminix (insecticide) tubes for tick
control, 236
DDT, 223–224, 227, 231
Deer, mule (*Odocoileus hemionus*), 62
Deer, white-tailed (*Odocoileus virginianus*):
"4-poster" device for, 244; acaracide
application to, 244–245; adult ticks and,
240; antitick vaccine for, 245; contracep-
tives for, 243; fences for exclusion of, 243;
geographic distribution of, 61–62;
increased populations of, 61–63; removal
or reduction of, 241–243
Deer tick (blacklegged tick). See *Ixodes
scapularis*
Deer tick virus encephalitis: agent of, 210;
diagnosis of, 211; geographic distribution
of, 210; incidence of, 205; manifestations
of, 209–210; prevalence in ticks, 210;
reservoirs of, 210; treatment of, 211;
vaccine for, lack of, 212
DEET (*N,N*-diethyl-*m*-toluamide), 257–258
Depression and Lyme disease, 33, 97, 99
Dermacentor variabilis (American dog tick),
213, 235, 249, 259–260
Diagnosis: general approach to, 87–91; risk
factor assessment for, 91–92; time pattern
of symptoms for, 93
Diarrhea from antibiotics, 163–164, 179
Direct detection of *B. burgdorferi*:
cultivation for, 119–121; microscopy for,
121; polymerase chain reaction for (*see*
PCR test)
Dogs: diagnosis in, 174; flea-and-tick collars
for, 259–260; incidence of, 40, 174;
kidney disease with, 41, 174–175;
manifestations of, 40–41, 174; oral drugs
for tick control for, 260; spot-on
insecticides for, 260; tick check for, 259;
treatment of, 174–175; vaccines for,
264–269

Dog tick, American. See *Dermacentor
variabilis*
Do-it-yourself (self) diagnosis, 97–99,
138–139
Doxycycline, 140, 153, 155–157, 174–175, 178,
189, 191, 202–203, 240. *See also*
Tetracycline antibiotics

EIA (enzyme-linked immunoassay) test:
antibiotics, effect on, 111, 115; false-
positive results from, 105, 110–111;
improvements in, 108–109; index
calculation for, 108; interpretation of
results from, 109–111, 131, 135–136;
method for, 106–108; strain or species
differences and, 105, 132–133
ELISA test. *See* EIA test
Environmental Protection Agency (EPA),
226, 228, 237, 246, 256n, 258
Epidemiology of Lyme disease, 78–81
Erythema migrans: absence of, during Lyme
disease, 18, 24, 102; acrodermatitis
chronica atrophicans and, 96; appearance
of, 17–18, 95; biopsy of, 119–120, 154; case
report definition of, 79; cellulitis versus,
94, 180; duration of, 95; history of, 1–4; late
Lyme disease and, 95; location of, 18;
multiple rashes of, 20; pictures of, 19–20;
progression of, 94–95; seasonality of, 94;
size of, 18, 79; tick bite allergy versus, 95;
treatment of, 154–157, 178–179
Erythromycin, 154. *See also* Macrolide
antibiotics
Eyes, infection of, 38

False-negative and false-positive laboratory
tests, 109–111
Fatality, risk of, from Lyme disease, 37–38
Fatigue, symptom of: chronic fatigue
syndrome and, 185–186; cytokines and,

17; early Lyme disease and, 3, 7, 95, 156; fibromyalgia and, 186–187; frequency in general population of, 195; non-specificity of, 99–100, 135; post-infection syndrome and, 185; post–Lyme disease treatment syndrome and, 156, 184

Fences for deer control, 222, 242–243

Fever: babesiosis and, 203, 205; coinfections and, 215; deer tick virus encephalitis and, 209, 211; early disease and, 3, 4, 7, 23, 96–97, 120, 167; human granulocytic anaplasmosis and, 199–202; Jarisch-Herxheimer reaction and, 167–168; relapsing fever and, 212; Southern tick-associated rash illness and, 217

Fibromyalgia, 182–183, 185–187

Fipronil insecticide, 228, 260

Flagyl. See Metronidazole

Flea and tick collars for pets, 228–230, 259

Fluoroquinolone antibiotics, 148, 161

Food and Drug Administration (FDA): antibiotics and, 153, 160–162, 191; diagnostic tests and, 109, 114, 123, 127, 129–130, 207; insecticides and, 230; vaccines and, 263–267

Grunwaldt, Edgar, 126

Heart, infection of: diagnosis of, 28; fatalities with, 37–38; frequency of, 37; manifestations of, 27–28, 36–38; treatment of, 28, 176, 179

Hives, 95, 166

HIV infection and Lyme disease, 25–26

Horses, 41, 268

Human granulocytic anaplasmosis (HGA): cause of, 199; diagnosis of, 201–203; incidence of, 199–200; manifestations of, 200–201; mortality from, 200; reservoirs for, 200; treatment of, 202–203, 253

Hydroxychloroquine for Lyme arthritis, 176

Hyperbaric oxygen therapy, 175–176

IFA (indirect immunofluorescence assay), 124–125

IgM antibody tests, 115–116

Imidacloprid insecticide, 228, 259

Immunity: adaptive, 21–22, 48, 194; antibiotic therapy's effect on development of, 111, 115; cytokines and, 16–17, 34; innate, 16, 21–22, 43, 48, 194; lymphocytes and, 22, 24

Immunoblot test. See Western blot assay

Immunodeficiency, tickborne disease risk with, 25

Infectious Diseases Society of America, guidelines of, 177–178, 272

Innate immunity. See Immunity: innate

Insecticides: environmental effects of, 223–224, 227–229; types of, 227. See also Acaricides

Integrated pest management for tick control, 246

International Lyme and Associated Diseases Society (ILADS), 9, 178

Internet resources, 271–273

Intravenous antibiotic therapy: complications of, 140, 161–162, 164–165; indications for, 179; methods of, 164; oral antibiotic therapy, comparison with, 155

Ivermectin, 244

Ixodes pacificus: Borrelia bissettii, carriage of, by, 54; frequency of infection of, 78; geographic distribution of, 69, 78; lizard blood in, 54

Ixodes persulcatus (taiga tick), 55, 200, 210, 213

Ixodes ricinus (sheep tick), 1, 54–55, 75, 200, 210, 213, 252, 256, 260

Ixodes scapularis (blacklegged tick): antibiotic prophylaxis after bite by, 252–254; geographic distribution of, 54; *Ixodes dammini*, similarity to, 3, 54; northern and southern forms of, 54; picture of, 20; vaccine against, 239, 245, 264
Ixodes ticks, other species, 55

Jarisch-Herxheimer reaction, 167–168
Joint (synovial) fluid, analysis of, 122, 194
Joints, infection of: autoimmunity in, 35–36, 195; laboratory diagnosis of, 122, 194; manifestations of, 2–3, 27, 33–35; treatment of, 176, 179, 194–195

Laboratory tests: cultivation of spirochetes, 119–121; direct detection (*see* Direct detection of *B. burgdorferi*); Do-it-yourself (DIY) ordering of, 138–139; EIA (*see* EIA test); FDA approval for, 129–130; indirect immunofluorescence assay (IFA) (*see* IFA); lymphocyte proliferation test (*see* Lymphocyte); polymerase chain reaction (PCR) test (*see* PCR test); predictive value of, 134–135; quality control, 128; reproducibility of, 130–131; sensitivity of, 134; specificity of, 135; urine antigen assay (*see* Urine antigen test for Lyme disease); Western blot (immunoblot) test (*see* Western blot assay)
Landscapes and lawns: disease risk and, 56, 62, 70, 76; modification of, for tick control, 229, 261
Larva (larval tick), 51–52, 56, 58, 77
Liver effects, 39, 169
Lone star tick. See *Amblyomma americanum*
Lumbar puncture. *See* Cerebrospinal fluid (CSF), analysis of

Lyme, Connecticut, 2–3, 34, 61, 141, 155, 221
Lyme arthritis. *See* Joints, infection of
Lyme borreliosis, 10–11
Lyme disease: antibiotic trials for, 151–157; case report definition for, 79–80; distribution of, 73–75, 91–92; epidemiology of, 78–81; fatality, risk of, 37–38; heart involvement in (*see* Heart, infection of); high risk areas for, 74, 81; history of, 1–4, 124–127, 263–264; incidence of, 63–65, 80–81; joint involvement in (*see* Joints, infection of); name origin for, 11; nervous system involvement in (*see* Nervous system, infection of); other symptoms and complications of, 23, 38–39, 96–97; risk factors for, 75–78, 91–92; skin involvement in (*see* Acrodermatitis chronica atrophicans; Erythema migrans); spread of, 63–65, 71–75; stages of, 30, 46, 93, 108, 170, 173–174; treatment recommendations for, 177–181; underreporting of, 80–81
"Lyme literate" doctor, 180
Lyme neuroborreliosis. *See* Nervous system, infection of
LYMErix vaccine, 263–265
Lymph nodes (glands), 23, 96, 186
Lymphocyte: proliferation assay of, for diagnosis, 133–134; types of, 22
Lymphocytoma, 96

Macrolide antibiotics: effectiveness of, against *Borrelia*, 147, 150, 154; infectiveness of, for HGA, 202; mechanism of action of, 142, 144, 147; non-antimicrobial effects of, 191; types of, 147. *See also under name of specific antibiotic*
Malawista, Stephen, 3, 151
Masters, Edward, ix, 217

Masters Disease. *See* Southern tick-associated rash illness
Memory and thinking, impairment of, 33, 82, 87, 97, 184, 186, 192, 211
Meningitis, 31
Metronidazole (Flagyl), 148–149, 164
Microbiome, effects of antibiotics on, 145, 163–164, 179, 189n
Monhegan Island, Maine, 235, 241–242
Mononucleosis, infectious, 111, 193
Mosquitoes: control of, 222–225, 234, 248, 260; differences of, from ticks, 49, 53, 204, 249; lack of transmission by, of Lyme disease, 53, 220; repellents for, 254–255, 257
Multiple sclerosis, 29, 33, 40, 99
Myalgic encephalomyelitis. *See* Chronic fatigue syndrome

Nantucket Island, Massachusetts, 60–63, 81, 205, 250
National Institutes of Health (NIH), 2, 221n
Nervous system, infection of: Bell's palsy (*see* Bell's palsy); cranial nerve involvement with, 32; diagnosis of, 117–118, 120, 179; encephalomyelitis with, 33; facial paralysis (*see* Bell's palsy); frequency of, 31; intravenous treatment of, 179; laboratory testing for, 116–118; manifestations of, 27, 31–33; meningitis with, 31; mental deficits with, 33, 97; nerve inflammation with, 32; oral treatment of, 178–179
Nondisease, phenomenon of, 99
Nymph (nymphal tick), 20, 51–52, 54, 65–69, 71, 77–78, 198, 205, 213, 215

Odocoileus virginianus. See Deer, white-tailed
Oil of lemon eucalyptus. *See* Repellents: PMD
Open access journals, reliability of, 9

Oral antibiotic therapy: dosing of, 145, 169, 179; indications for, 178–179
Organophosphate insecticides, 227, 230
OspA protein: vaccine, based on, 263–266; Western blot test and, 128

PCR (polymerase chain reaction) test, 122–123
Penicillin antibiotics: allergic reactions to, 166; characteristics of, 144–145, 147; effectiveness of, against *Borrelia*, 147, 150–152; ineffectiveness of, for HGA, 202; treatment with, 152. *See also under name of specific antibiotic*
Permethrin. *See* Pyrethrins and pyrethroids for tick control
Peromyscus leucopus (white-footed mouse): difference from laboratory mice, 56n; ecology of, 56, 66; geographic distribution of, 56, 71; infection of, 56; removal of, for tick control, 235–236; vaccine targeting of, 237–239
Personal and household protection: clothing for, 249–250; landscape modification for, 229, 261; repellents for (*see* Repellents); tick check for, 250; tick removal for, 250–252
Pesticides. *See* Acaricides; Insecticides
Pets. *See* Cats; Dogs; Horses
Phagocytes (phagocytic white cells), 16, 199–200
Pheromones for tick control, 234–235
Placebo, 24, 151–152, 187–189, 253, 263
Plaquenil. *See* Hydroxychloroquine for Lyme arthritis
Post-infection syndromes, other, 185
Post–Lyme disease treatment syndrome (PLDTS): characteristics of, 150, 182–185; conjectures about causes of, 192–195; cytokines and, 192–193; remnant DNA

Post–Lyme disease (*continued*)
or protein and, 194–195; treatment trials for, 150, 187–189; unintended effect of antibiotics and, 189–192

Powassan virus encephalitis, 209n

Prednisone. *See* Corticosteroids and Lyme disease treatment

Pregnancy, Lyme disease during, 25

Pyrethrins and pyrethroids for tick control: area application of, 229–230, 236; on clothing, 257; flea-and-tick collars and, 259; mechanism of action, 229; toxicity from, 230; types of, 229

Qualities, primary and secondary, 85–87, 168, 185

Quinine for babesiosis, 208

Raccoons, 57, 205, 237–238, 240

Relapsing fever, 45, 212–213. See also *Borrelia miyamotoi*, infection with

Repellents: 2-undecanone, 259; characteristics of, 254–255; DEET (*N,N*-diethyl-*m*-toluamide), 257–258; formulation of, 256; IR3535, 257–258; mechanism of action of, 254–255; picaridin (KBR3023), 257–258; PMD (*p*-methane-3,8-diol), 257, 259; precautions with, 258; testing of, 255–256

Reservoirs: birds as, 57–58; chipmunks as, 57; deer as, 58; dogs as, 59; *Peromyscus leucopus* as (see *Peromyscus leucopus*); rats as, 57, 235–236; shrews as, 57; squirrels as, 57; voles as, 57, 67, 204–205, 236–237; white-footed mouse as (see *Peromyscus leucopus*)

Rocephin. *See* Ceftriaxone

Rocky Mountain Laboratories, 2

Screening for disease, 103–104

Seasons and Lyme disease risk, 77, 94

Selamectin, 260

Sensitivity of laboratory tests, 134

Seronegative Lyme disease, 136–138

Sevin. *See* Carbaryl

Sheep tick. *See Ixodes ricinus*

Shelter Island, New York, 2, 4, 126

Shrews, 57, 67, 200, 205, 237, 239

Side effects. *See* Antibiotics: adverse (side) effects of

Silver, colloidal, therapy with, 175

[S]-methoprene, 224, 260

Southern tick-associated rash illness (STARI): geographic distribution of, 74, 217; manifestations of, 217; treatment of, 217. See also *Amblyomma americanum*

Specificity of laboratory tests, 134

Spinal tap. *See* Cerebrospinal fluid (CSF), analysis of

Spinosad insecticide, 260

Spirochetes, biology of, 42–44

Squirrels, 57, 237, 261

Staphylococcus aureus (staph bacteria, MRSA), 15, 78, 147, 180

Steere, Allen, ix, 3, 126, 151

Strains of *B. burgdorferi*: definitions for, 46; differences in invasiveness of, 23; geographic distributions of, 4–7; immune response differences for, 56, 132–133

Sulfa antibiotics, 148

Symptoms and signs of Lyme disease: in general, 85–87; time pattern for, 92–93; unusual, during Lyme disease, 92. *See also under name of different organ system*

Syphilis, comparison of, with Lyme disease, 30, 44–45

Taiga tick. See *Ixodes persulcatus*

Tetracycline antibiotics: adverse effects of, 156, 168; effectiveness of, against *Borrelia*, 147, 150, 152, 155–157; mechanism of action, 142, 144; non-antimicrobial effects of, 190–191; types of, 148. *See also under name of specific antibiotic*

Tick bites: antibiotic prophylaxis after, 252–254; other infections from (*see* Cellulitis); reactions to, 14, 95

Tick-borne encephalitis (TBE) virus: diagnosis of, 211; geographic distribution of, 210; incidence of, 210; manifestations of, 210–211; treatment of, 211; vaccine for prevention of, 211–212

Ticks: avoidance of, 248–250; blood meal for, 52–53; classification of, 50–51, 213; identification of, 249, 252; life cycle of, 51–52; removal of, from skin, 250–252; types of, 53–54, 249

Tigecycline, 148, 161

Transmission of Lyme disease: lack of, by person-to-person contact, 3, 45n, 76; pregnancy and risk of, 25; by ticks, 14, 75–76; transfusion and risk of, 76, 205

Twin Earth parable, 5–10, 177–178

Two-tier testing, 112, 131, 134

Tygacil. *See* Tigecycline

United States Department of Agriculture (USDA), 265–266

Urine antigen test for Lyme disease (LUAT), 133

Vaccines: against ticks, 245, 264; bacterin-based, 265–267; for dogs, 264–268; for humans, 262–264; OspA-based, 265–266; for wildlife, 237–239

Vancomycin, 150, 164

Vector control agencies, 224

Vibramycin. *See* Doxycycline

VlsE protein, 108–109, 114

Voles, 57, 67, 204–205, 236–237

Western blacklegged tick. See *Ixodes pacificus*

Western blot (immunoblot) assay: criteria for, 114, 127–128; EIA, difference from, 113; false-positive results from, 105; history of, 125–127; interpretation of results from, 113–114; method for, 112–113; standardization of, 127

White cell count (WBC), 22

White-footed mouse. See *Peromyscus leucopus*

Zithromax. *See* Azithromycin